REFRACTORY METALS

METALS

State-of-the-art 1988

REFRACTORY METALS

State-of-the-art 1988

Proceedings of a symposium held on Refractory Metals: State-of-the-Art, sponsored by the TMS Refractory Metals Committee, held during the TMS Fall Meeting, Chicago, Illinois, September 27, 1988.

Edited by

P. Kumar
Cabot Corporation
Boyertown, PA

and

R.L. Ammon
Westinghouse Adv. Energy Systems
Pittsburgh, PA

A Publication of
TMS
Minerals • Metals • Materials

A Publication of The Minerals, Metals & Materials Society
420 Commonwealth Drive
Warrendale, Pennsylvania 15086
(412) 776-9024

Printed in the United States of America
Library of Congress Catalog Number 89-60380
ISBN Number 0-87339-088-1

Preface

A symposium on "Refractory Metals: State-of-the-Art" was held during the Metals Congress in October 1988. It was sponsored by the Refractory Metals Committee of TMS. Most of the papers which were presented in this symposium are compiled in this book. About half of the papers are related to extra-terrestrial transportation and power systems; the remaining half pertain to the processing of refractory metals.

Titran, Stephens and Petrasek's paper reviews the on-going research at NASA Lewis Research Center. Niobium alloys and tungsten-based composites are being evaluated for nuclear power systems for extra-terrestrial space. These systems are expected to operate above 1300°K with advanced systems being designed for 1700°K. A new fabrication process, arc-spray deposition, for composites is described.

Grobstein used this technique to evaluate Nb alloy-W fiber composites. The orientation of fiber relative to the stress axis has strong affect on the creep resistance of composites. Composites are superior to monolithic niobium alloy of equivalent composition.

Properties of powder-metallurgy and ingot-metallurgy niobium alloy WC-3009 (Nb-30Hf-9W) are compared by Hebsur and Titran. While the tensile strength of both alloys are similar, P/M-processed alloy exhibited lower ductility. The processing of P/M alloy was relatively easy.

Effect of HfC additions on tensile behavior of tungsten was also investigated by Yun. The presence of a dispersoid results in superior tensile properties. HfC is better than ThO2 or potassium bubble due to higher orowanian stresses.

Chen et al. have also examined the effect of HfC additions on properties of W-Re alloys; these alloys are being considered for high-power space systems. The elevated temperature tensile strength, creep strength and usable temperature of W-Re alloys increases by the presence of fine HfC particles.

Tsao et al. have investigated W-Re system with substantially higher Re content than in Chen's investigation. Addition of thoria increases both the ductility and strength at elevated temperatures. These are explained on the basis of increase in twin deformation and retarded grain boundary sliding.

Desai et al. discuss the corrosion resistance of several refractory metals in molten lithium fluoride. Lithium fluoride might be used for thermal storage in advanced solar dynamic space power systems. The corrosion resistance decrease in the order of Mo, W, Nb, Ta and Zr. Nb-Zr alloys have a higher corrosion rate than Nb; removal of moisture and addition of Zr to salt improve the corrosion resistance of Nb Zr alloy.

Jackson et al. produced refractory metal preforms, including tubular shapes, by low pressure plasma deposition. Mechanical tests after subsequent thermomechanical processing indicate good strength, but poor ductility. It is explained on the basis of high interstitial contents. The need for refractory metal powder with low interstitial contents is obvious.

The rapid solidification of niobium alloys is discussed by Jha and Ray. The ultra-fine microstructures are obtained by this technique. This results in higher strengths at low temperatures. However, the high temperature strength of the rapidly solidified alloys fell rapidly and is comparable to commercial alloys.

Robino discusses an interesting concept for improving the weldability of powder metallurgy grade molybdenum. It is possible to reduce the center line cracking and pore formation in the P/M molybdenum by incorporating titanium or hafnium in joints prior to welding. Mechanisms by which these additions improve the welding are discussed. It may be possible to use similar concept for welding other crack-prone refractory metals.

An interesting and novel method (fused salt electrolysis) for producing refractory metals is discussed by Sadoway. This method is in developmental stages. Its potential for synthesizing advanced materials in assessed. Its applications to the extraction, refining and plating of refractory metals are described.

Lipschutz et al. investigated the supercooling of Nb-Si system using electromagnetic levitating processing. The solidification sequence is described as a function of the supercooling of the melt. It is possible to obtain metastable and amorphous phases by the supercooling. The zone of coupled growth of eutectic is feasible.

The absorption and desorption behavior of nitrogen in liquid niobium was investigated by Park using an electromagnetic levitation technique. The nitrogen dissolution in liquid niobium is exothermic. The rate-controlling step for absorption is either the adsorption of nitrogen molecules on the liquid surface or the dissociation of molecular nitrogen into atomic nitrogen.

I would like to thank the authors for sending manuscripts in on time. Mary Kretzman and Catherine Yoder have provided invaluable secretarial assistance.

P. Kumar

Table of Contents

REFRACTORY METAL ALLOYS AND COMPOSITES FOR SPACE NUCLEAR POWER SYSTEMS

Robert H. Titran, Joseph R. Stephens, and Donald W. Petrasek

National Aeronautics and Space Administration
Lewis Research Center
Cleveland, Ohio 44135

Summary

Space power requirements for future NASA and other United States missions will range from a few kilowatts to megawatts of electricity. Maximum efficiency is a key goal of any power system in order to minimize weight and size so that the space shuttle may be used a minimum number of times to put the power supply into orbit. Nuclear power has been identified as the primary power source to meet these high levels of electrical demand. One method to achieve maximum efficiency is to operate the power supply, energy conversion system, and related components at relatively high temperatures. For systems now in the planning stages, design temperatures range from 1300 K for the immediate future to as high as 1700 K for the advanced systems. NASA Lewis Research Center has undertaken a research program on advanced technology of refractory metal alloys and composites that will provide base line information for space power systems in the 1900's and the 21st century. Special emphasis is focused on the refractory metal alloys of niobium and on the refractory metal composites which utilize tungsten alloy wires for reinforcement. Basic research on the creep and creep-rupture properties of wires, matrices, and composites will be discussed.

Refractory Metals: State-of-the-Art 1988
Edited by P. Kumar and R.L. Ammon
The Minerals, Metals & Materials Society, 1989

Introduction

The objective of our research on refractory metals is to provide an understanding of their behavior and capabilities under conditions that simulate advanced space power system requirements. Current research is focused on monolithic materials to identify alloys that may meet the demands of near term space power components. In addition, refractory metal alloys are being considered for matrices and fibers to be used in metal matrix composites developed for more long term needs that will have to be met in the late 1900's or in the 21st century. These alloys and composites are anticipated to be used in heat generation systems in such applications as cladding for nuclear fuel pins, for heat pipes and tubing, and in energy conversion systems such as the Stirling engine for heater heads, regenerators, pressure vessels, and heat pipes.

The research activities underway on the refractory metals are conducted primarily in-house with some supporting research being done on university grants. The purpose of this paper is to present a summary of our research activities currently underway and to briefly describe the future direction of our research.

Space Power Materials Needs

Current spacecraft require electrical power in the few hundred watts to about 75 kWe range as shown in Fig. 1 (1). The former NASA Skylab operated with a little over 10 kWe of electrical power. In contrast, the space station, NASA's next major space system is anticipated to require nearly 100 kWe initially and to grow to meet ever increasing demands to several hundred kilowatts of electric. Future missions being planned under the Civil Space Technology Initiative (CSTI) and Pathfinder, which include a lunar base or a manned flight to Mars, are expected to push the power requirements to even higher levels as shown in the figure such that tens to hundreds of megawatts of electrical power will be needed.

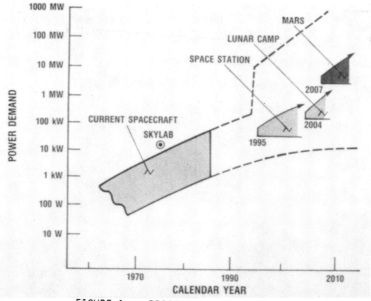

FIGURE 1. - PROJECTED GROWTH IN SPACE POWER.

Possible sources for the power levels described previously are shown in Fig. 2 (2). For lives of over a year and power levels of less than 10 kWe, solar or radioisotope power sources have been used by the United States. In 1965, the United States successfully launched SNAP-10A, the first nuclear reactor to be operated in space. Since then the Soviet Union has used reactors routinely in low, short-term orbits. It can be seen from this figure, that for the 7- to 10-year lives and power requirements anticipated for future systems, nuclear power is the only power source that can be considered.

A joint NASA, DOD, and DOE program to develop a space nuclear reactor capability is currently under way called SP-100 (3), envisioned as a 100 kWe Space Power nuclear reactor; hence, SP-100. The program has focused on demonstrating a Ground Engineering System (GES). This program is managed by the Jet Propulsion Laboratory of NASA and a contract has been awarded to GE to design and build the GES. It is anticipated that the SP-100 type of reactor will be able to support a broad spectrum of space activities that will require large amounts of electrical power including communications, navigation, surveillance, and materials processing. Some of the material related constraints for SP-100 are listed in Fig. 3 and include the use of liquid lithium reactor coolant, 1350 K, 7-year, 1-percent strain design criteria, and a 3000-kg system weight that can be launched by the space shuttle. Based on these constraints, we have undertaken an advanced technology program to allow for future growth of the current power level envisioned for SP-100.

Another high temperature materials need is in the energy conversion system for the high electrical power levels that are being considered. Both the Brayton and Stirling systems are under consideration to convert the heat energy, whether solar or nuclear, into electrical energy (4-5). NASA Lewis has paved the way for Stirling engine technology both for terrestrial and

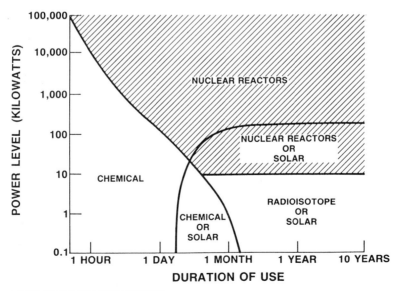

SP100 SPACE REACTOR SAFETY
DOE/NE0083 MAY 1987

FIGURE 2. - POWER SOURCES FOR SPACE APPLICATIONS.

MATERIAL CONSTRAINTS
- **LIQUID LITHIUM REACTOR COOLANT**
- **1350 K-7 YEAR-1% STRAIN DESIGN CRITERIA**
- **3000 kg SYSTEM WEIGHT**

FIGURE 3. - SP-100 GOAL: TO DEVELOP NUCLEAR POWER FOR SPACE.

space applications. Shown in Fig. 4 is a cross section of a free-piston space power demonstrator engine that will deliver 25 kWe (6). This engine operates at 1050 K; a lower temperature than the 1350 K that is anticipated for SP-100 and therefore uses a molten salt as the heat source for ground demonstration purposes and is constructed of stainless steels, superalloys, and other materials suitable for low temperature applications. For space applications, refractory metals will be required in such areas as the pressure vessel, heat pipes, heater head, regenerator, and other structural members because of the higher material temperature requirements. The material constraints shown previously in Fig. 3 will have to be met by the refractory metals which will probably be niobium or molybdenum base alloys or composites.

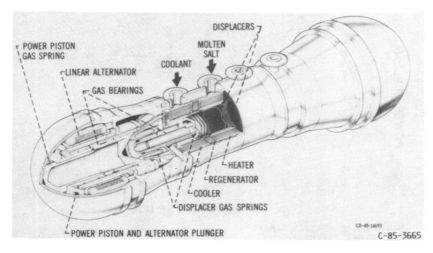

FIGURE 4. - SPACE POWER DEMONSTRATOR ENGINE.

Refractory Metal Alloy Technology

The SP-100 program has selected Nb-1Zr as the nuclear assembly test
NAT) material for the GES. NASA Lewis in concert with Oak Ridge National
aboratory (ORNL) and Westinghouse-Advanced Engineering System Division
W-AESD) is conducting a series of studies to define the creep behavior of
his alloy so that design engineers will be able to use this alloy with con-
idence.

It has been determined from reiterative design considerations that
lb-1Zr has marginal strength for the SP-100 GES. To improve its 7-year
350 K 1-percent creep strength, such things as a large grain size micro-
tructure (80 μm) are being explored. Also heats with increased tantalum
.nd tungsten contents (but still within the allowable specifications) are
inder test. An alternate approach that we are exploring is to consider a
igher strength alloy that will still meet the material constraints for
.P-100. We have selected an alloy called PWC-11 with a composition of
lb-1.0Zr-0.1C that was developed in the 1960's. This alloy has been previ-
iusly tested in lithium and has been shown to be compatible under conditions
.nticipated for SP-100. However, several question remained to be answered
'rom earlier studies. For example, the weldability of PWC-11 and the subse-
iuent effects on creep properties are not fully characterized, and the long
'erm stability of the carbon precipitates (which are believed to be respon-
ible for the improved strength of this alloy) is not known. We currently
ave creep tests underway at low stress levels similar to those that may be
ncountered in SP-100 fuel pin claddings. Both NB-1Zr and PWC-11 are being
ested at 1350 K and a stress of 10 MPa. With tests in excess of 20 000 hr,
s shown in Fig. 5 PWC-11 has not achieved any practical measurable creep
eformation while Nb-1Zr has reached 1-percent creep in 11 000 hr and
-percent creep in 18 800 hr. The results to date clearly demonstrate the
uperiority of PWC-11.

The microstructures of Nb-1Zr and PWC-11 are compared in Fig. 6. After
he standard 1-hr, 1475-K anneal, Nb-1Zr has an average grain size of 20 μm

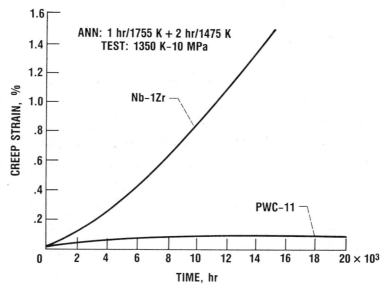

FIGURE 5. - CREEP CURVES FOR Nb-1Zr AND PWC-11.

with a few precipitates present, which are believed to be ZrO$_2$. The microstructure of PWC-11, which had undergone its standard anneal of 1 hr at
1775 K plus 2 hr at 1475 K, is shown in Fig. 6. The annealed material had a
mixture of elongated grains with an average grain size of 25 μm measured by
the circle-intercept method, with an aspect ratio of approximately 5:1.
Numerous shapes and sizes of particles were apparent in the microstructure.
The morphology ranges from massive 5 μm particles to submicron needle-like
particles which appear to be oriented on slip planes as has been reported
previously (7). The majority of particles are believed to be primary carbides which formed during the initial solidification and are neither broken
up during the sheet rolling process nor dissolved during the annealing process. A typical microstructure of an EB welded PWC-11 test specimen is
shown in Fig. 7. The base metal away from the weld is in the annealed condition. The weldment has a columnar structure with the grain size ranging
from about 45 to over 200 μm. The weld zone exhibits extensive second phase
precipitation similar to the annealed condition except that the particles
appear to be finer and form cell-like domains within the grains.

The high temperature creep strength of PWC-11 (>0.5 T$_m$), relative to
the order of magnitude lower carbon content Nb-1Zr alloy as shown in Fig. 5,
has been attributed to the presence of very fine precipitates of (Nb,Zr)$_2$C
and/or (Nb,Zr)C ranging in size from 1 to 10 μm in diameter (8). As with
all precipitation-strengthened alloys, the long term beneficial contribution
of the precipitate to high creep strength is suspect. It has been postulated that welding and/or isothermal aging of the PWC-11 alloy could result
in a significant loss (>50 percent) in elevated temperature creep strength
(9). To verify or disprove this postulation, we conducted creep tests in
high vacuum (10^{-7} Pa) at 1350 K and 40 MPa to assess the effects of EB welding on creep strength. The creep curves to approximately 1-percent strain
are shown for Nb-1Zr and PWC-11 in Fig. 8. The PWC-11 annealed condition
(clearly the most creep resistant state) required about 3500-hr to 1-percent
strain. A similarly treated sample with an EB weldment required 2125 hr,
about a 30-percent decrease in the time for 1-percent strain. A Nb-1Zr
specimen was tested in creep for comparison to the annealed PWC-11 alloy.
As shown in Fig. 8, the time for 1-percent creep required about 75 hr, a
factor of about 45 compared to the annealed PWC-11 and a factor of about 28
for the welded condition. It should be noted that Nb-1Zr was annealed at
1775 K for this comparison which resulted in a grain size of about 45 μm.
This larger grain size compared to only 25 μm for PWC-11 should favor a
higher creep strength for Nb-1Zr.

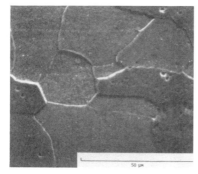

Nb-1Zr
ANNEALED 1 HR-1475 K

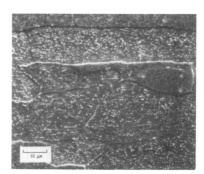

PWC-11   Nb-1Zr-.1C
ANNEALED 1 HR-1755 K + 2 HRS-1475 K

FIGURE 6. - MICROSTRUCTURES OF Nb-1Zr AND PWC-11.

FACE

ROOT

CROSS-SECTION OF SINGLE-PASS FULL PENETRATION WELD

UNAFFECTED BASE METAL

WELD FUSION ZONE

FIGURE 7. - MICROSTRUCTURES OF ELECTRON BEAM
WELDED PWC-11.

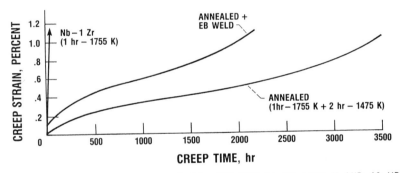

FIGURE 8. - CREEP CURVES OF Nb-1Zr AND PWC-11 AT 1350 K AND 40 MPa.

Moore et al. conducted short-time creep rupture tests to further characterize the effects of EB welding on PWC-11 (10). The EB welds in these
tests were perpendicular to the test axis. Tests were conducted in a
10^{-5} Pa vacuum at 1350 K after the post-weld heat treatment (1 hr at 1475 K)
and after aging at 1350 K for 1000 hr. In all the creep rupture tests of
these specimens, failure occurred in the unaffected base metal (Fig. 9),
thus demonstrating that the weld region was stronger.

Based on our creep tests conducted to date, projections have been made
for the stress for 1-percent creep in a 7-year time frame and compared to
the design requirements for SP-100. The results are shown in Fig. 10.
PWC-11 is a factor four times stronger than Nb-1Zr (20 to 5 MPa at 1350 K)
over the SP-100 design temperature range of 1350 to 1380 K and affords
excellent growth potential over the present SP-100 design stress criterion.

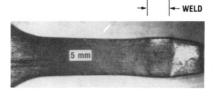

FIGURE 9. - TYPICAL BASE METAL CREEP
RUPTURE FAILURE IN EB WELDED PWC-11
MATERIAL TESTED AT 1350 K.

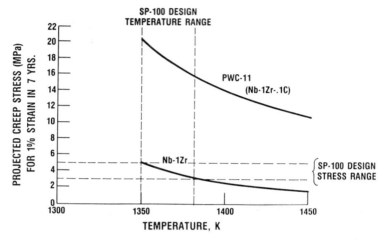

FIGURE 10. - CREEP POTENTIAL OF PWC-11 COMPARED TO Nb-1Zr.

The strength advantage of PWC-11 over Nb-1Zr was explored in more
detail by Grobstein and Titran (7). In particular, the concern about over-
aging of the precipitates during high-temperature exposure for long times
was addressed by microstructural characterization of the precipitate (complex carbides) morphology. Several techniques were used including light
metallography, scanning and transmission electron microscopy, x-ray diffraction, and chemical analysis of extracted particles. Table I summarizes the
results of this study. In the as-rolled condition, the precipitates were
relatively coarse, 1 to 10 μm in size, and were found to be hcp Nb_2C. After
an initial heat treatment of 1 hr at 1755 K and 2 hr at 1475 K, a different
finer precipitate formed. These particles were 0.05 to 0.1 μm in diameter
and were determined to be fcc (Zr,Nb)C with the Zr/Nb ratio approximately
70:30. After approximately 5000 hr at 1350 K (0.5 T_m), these fine precipitates almost doubled in size, but did not "overage" and were still effective
in pinning dislocations and thus resisting plastic deformation in creep.

TABLE I. – CHARACTERIZATION OF PRECIPITATES IN PWC-11

(Nb-1Zr-0.1C)[a]

	As-rolled	After initial heat treatment	After long-term high-temperature exposure
Size, μm	1 to 10	0.05 to 0.1	0.1 to 0.15
Structure	HCP	FCC	FCC
Composition	Nb_2C	(Zr,Nb)C	(Zr,Nb)C

[a]Conclusions: Aging at 1350 or 1400 K with an applied stress does not "overage" the precipitates. After long times (5000 hr) at 1350 K, the precipitates are still effective at pinning dislocations and resisting plastic deformation in creep.

Refractory Metal Composites Technology

The objective of this part of our program is to characterize wires, matrices, and composites for future space power systems where requirements for several hundred kilowatts to megawatts of electricity will need to be met. This advanced technology program focuses on tungsten fibers and Nb or Nb-1Zr matrices and thus can be compared directly with results from the SP-100 GES program on the Nb-1Zr and PWC-11 monolithic alloys. It is anticipated that these composites will enable the technology for advanced space power systems to be more efficient and provide more electrical power by allowing operation at higher temperatures.

Refractory metal alloys have been explored as potential fibers for a variety of matrices (11-12). The creep rupture strength is of primary significance for space power applications since the intended use of the material is for long time operations. A secondary consideration is the density of the fiber since there is a design weight criterion for launching into orbit. The ratio of the 100-hr rupture strength to density for a number of potential refractory metal wires is plotted in Fig. 11 for tests conducted at 1365 and 1480 K. It can be seen that for anticipated growth to higher temperatures, the tungsten-base and molybdenum-base alloys have the superior properties. The strongest alloys, W-Re-Hf-C and Mo-Hf-C are not commercially available so two lower strength tungsten compositions were selected for our initial studies since processing conditions and fiber-matrix interactions can be simulated directly with the tungsten base alloys. The two selected are 218 CS, an example of a lower strength, unalloyed lamp filament alloy and ST300 (W-1.5ThO$_2$), an example of a stronger, oxide dispersion strengthened alloy. Figure 12 is a plot of the time to rupture as function of stress at 1400 and 1500 K for these two alloys. It should be noted that at the longer times, 1000 to 10 000 hr, the two alloys' strength properties converge at both test temperatures. Based on these results we proceeded to fabricate composites using both compositions in the form of 0.20-mm-diameter wire as the reinforcement fiber material and Nb and Nb-1Zr as the matrix material.

The composites were fabricated using an arc-spray process developed at NASA Lewis which is shown schematically in Fig. 13 (13). In this process, the tungsten alloy fibers were wound on a drum using a lathe to accurately align and space them. The drum was inserted into a chamber which was subsequently evacuated and backfilled with argon. The Nb or Nb-1Zr matrix material in the form of 1.59-mm-diameter wire was arc sprayed onto the drum surface by using a pressurized argon gas stream. After spraying, the coated

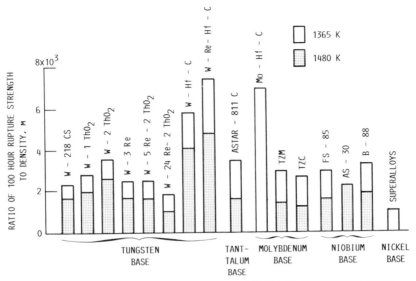

FIGURE 11. - STRENGTH COMPARISON OF CANDIDATE REFRACTORY METAL ALLOY FIBERS.

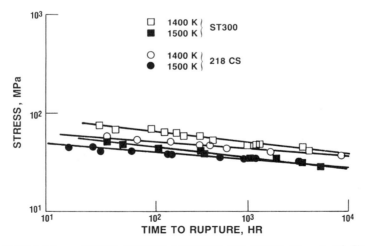

FIGURE 12. - STRESS RUPTURE STRENGTH FOR ST300 (W + 1.5% ThO$_2$) AND 218 CS FIBERS.

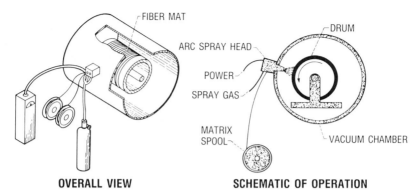

OVERALL VIEW   SCHEMATIC OF OPERATION

FIGURE 13. - ARC SPRAY MONOTAPE FABRICATION PROCESS.

fiber (monotape) was removed from the drum surface, cleaned, cut to size, stacked in three layers plus matrix only arc-sprayed monotapes on either surface, sealed in a container, and hot isostatically pressed (HIPed) to produce unidirectional fiber-oriented composites. HIP processing parameters were optimized for each combination of fiber and matrix to achieve the best possible properties of the composite. Parameters investigated included temperature, time, and pressure and were varied to explore under what conditions insufficient bonding between fiber and matrix occurred and where excessive reaction took place, as depicted schematically in Fig. 14. A typical microstructure of an as-processed ST300/Nb composite (Fig. 15) indicates minimal reaction between the fiber and matrix during the arc spray and HIP processes. One of the principal concerns in the use of composites for long-term, high-temperature applications is the degree of fiber-matrix interaction. Excessive fiber-matrix reaction could degrade the fiber and thus the composite properties. Figure 16 compares the reaction at the fiber-matrix interface that occurred in ST300/Nb-1Zr composites exposed for

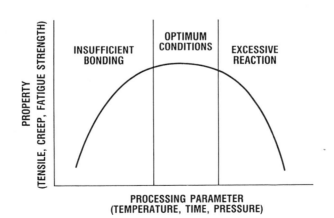

FIGURE 14. - OPTIMIZATION OF FABRICATION PARAMETERS
IS CRITICAL TO ACHIEVING MAXIMUM PROPERTIES.

11

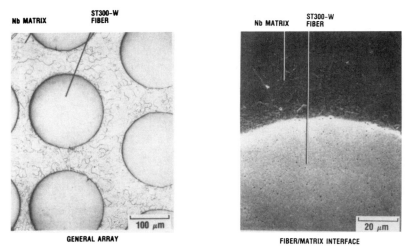

GENERAL ARRAY   **FIBER/MATRIX INTERFACE**

FIGURE 15. - TUNGSTEN FIBER REINFORCED NIOBIUM MATRIX COMPOSITES AS-FABRICATED MICROSTRUCTURE.

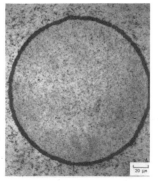 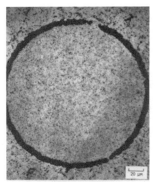

EXPOSED 1104 HRS AT 1400 K   **EXPOSED 921 HRS AT 1500 K**

FIGURE 16. - ST300/Nb + 1Zr FIBER/MATRIX REACTION.

about 1000 hr at 1400 and 1500 K. The depth of penetration into the 0.2-mm-diameter fiber is less than 0.01 mm. The effects of matrix composition on the depth of penetration are compared in Fig. 17 for ST300 fibers in Nb and Nb-1Zr matrices. After 2500-hr exposure the depth of penetration is about 0.01 mm for both matrices. The values for fiber-matrix reactions are in agreement with previously reported diffusion coefficients in the literature for tungsten-niobium diffusion couples (14). Our results thus indicate good microstructural stability for this composite system.

For the creep rupture investigation, tests were conducted on three-ply, unidirectional flat plates from which tensile specimens were cut by electrical discharge machining. An example of a specimen tested to rupture is shown in Fig. 18. Tungsten tabs were TIG welded on both sides of the ends of the composite specimens to prevent specimen shearing at the pin hole locations. It is also possible to make other shapes by the arc-spray, HIP

12

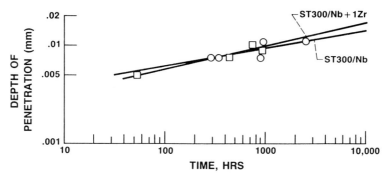

FIGURE 17. - DEPTH OF REACTION PENETRATION VERSUS TIME AT 1500 K.

FIGURE 18. - FRACTURED ST300/Nb + 1Zr COMPOSITE
CREEP RUPTURE SPECIMEN.

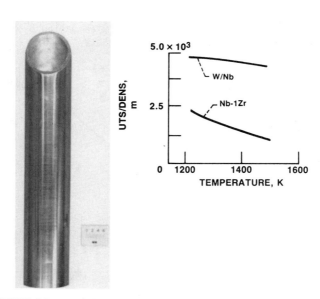

FIGURE 19. - W/Nb COMPOSITE TUBE AND COMPARISON OF
TENSILE STRENGTH/DENSITY RATIOS OF W/Nb WITH Nb-1Zr.

13

process, such as the tube illustrated in Fig. 19. Since for nuclear space power systems, fuel clads, along with heat pipes in tubular geometries will be required, the ability to produce composites having this geometry is very significant. Tensile strengths have been determined for the composites at temperatures of interest to advanced space power systems and are also shown in this figure (15).

The primary property requirement for space power system applications however, is adequate creep resistance for long time exposure. Preliminary creep results on the composite materials have been reported previously (16). A typical creep curve is reproduced in Fig. 20 for a ST300/Nb-1Zr composite tested at 1400 K under an applied stress of 180 MPa. The creep curves for these composites exhibit the characteristic three-stage creep behavior typical of tungsten and other metals and alloys at elevated temperatures. The strain to rupture ranges from 5 to 7 percent for the composite materials tested in this program.

The fracture surfaces of composite specimens were examined using scanning electron microscopy. Figure 21 shows the fracture surface of a ST300/Nb-1Zr composite where it should be noted that both the fiber and the matrix fail in a ductile manner in creep-rupture testing. Further evidence of fiber- and matrix-ductile behavior is shown in the optical micrograph of Fig. 22, where necking of the fiber and matrix can be observed.

The effective use of fiber reinforcements to increase the creep resistance of Nb and Nb-1Zr is shown in Fig. 23. The time to achieve 1-percent creep strain for arc-sprayed niobium under an applied stress of 20 MPa was 17 hr, while arc-sprayed niobium reinforced with 40-vol % ST300 fiber and stressed at an order of magnitude higher stress (200 MPa) has nearly an order of magnitude increase in the time to reach 1-percent strain. Increasing the fiber content results in further increases in creep resistance as shown for the 50-vol % fiber content ST300/Nb-1Zr composite.

Since the reinforcing fibers have a density over twice that of Nb, the 50-vol % composite is over one and a half times heavier than niobium, and thus density must be taken into consideration when making property comparisons. A comparison of the creep stress to density ratio for 1-percent

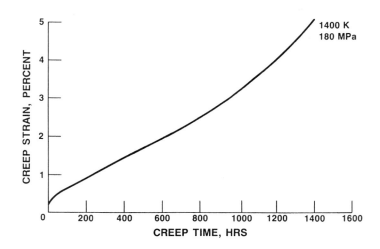

FIGURE 20. - TYPICAL CREEP CURVE FOR ST300/Nb + 1Zr COMPOSITE.

CREEP RUPTURE TEST DATA:

1500 K
150 MPa
284.5 HRS

200 µm

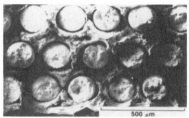

500 µm

FIGURE 21. - FRACTURE SURFACE OF
ST300/Nb COMPOSITE SPECIMEN.

CREEP RUPTURE
TEST DATA
1400 K
220 MPa
2325 HRS

100 µm

FIGURE 22. - FRACTURE SECTION
OF ST300/Nb + 1Zr CREEP
RUPTURE SPECIMEN.

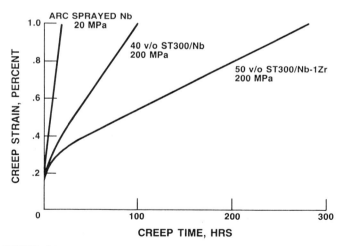

FIGURE 23. - 1400 K CREEP CURVES FOR ST300/Nb AND Nb-1Zr
COMPOSITES.

strain for the composites, PWC-11, and Nb-1Zr is made in Fig. 24. On this
density corrected basis, the composites are over an order of magnitude
stronger than Nb-1Zr and three and a half to four times stronger than PWC-11
at both test temperatures, 1400 and 1500 K.

15

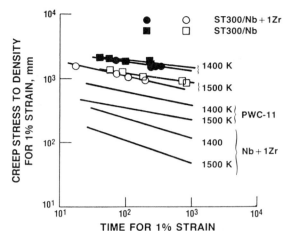

COMPOSITES NORMALIZED TO
50 VOLUME PERCENT FIBER CONTENT

FIGURE 24. - COMPARISON OF CREEP STRESS TO
DENSITY RATIO FOR 1 PERCENT STRAIN FOR
COMPOSITES, PWC-11 AND Nb + 1Zr.

A comparison of the minimum creep rates of the composites tested at 1400 and 1500 K with that for the arc-sprayed niobium monolithic material tested at 1400 K is made in Fig. 25. It is evident that the composites creep at a much lower rate than the niobium matrix material. Noting that the strain- and strain-rate compatibility must be maintained at the fiber-matrix interface during creep of a composite subjected to uniaxial loading, it is possible to estimate the relative magnitude of the stress on the matrix using Fig. 25. For example, it is evident from Fig. 25 that at 1400 K the ST300/Nb composites exhibit a minimum creep rate of about 1×10^{-8} sec^{-1} at 250 MPa. Using the strain-rate compatibility arguments, the data in Fig. 25 suggest that a stress of about 15 MPa would enable the niobium matrix to creep at the same rate. It can be shown using the rule of mixtures, that the corresponding stress on the matrix is only about 3 percent of the total applied stress acting on a composite containing 50-vol % fibers. This means a first order prediction of creep behavior of the composites can be described by the creep equations for the reinforcing fibers. The minimum creep rate of the composites can thus be equated to the power creep behavior as follows:

$$\dot{\varepsilon}_m = A \, \exp\left(\frac{-Q}{RT}\right)\sigma^n$$

$$\sigma = \sigma_f = \frac{\sigma_c}{V_f}$$

$$\dot{\varepsilon}_m = A \, \exp\left(\frac{-Q}{RT}\right)\left(\frac{\sigma_c}{V_f}\right)^n$$

where

σ_c the stress on the composite

σ_f the stress on the fiber assuming that the fiber carries the total load

16

V_f the volume-fraction-fiber content

Q the apparent activation energy

n the creep-rate stress exponent

A a constant for the fiber

The calculated composite creep activation energy Q of 465 to 490 kJ/mol agrees with results for other forms of tungsten tested in this temperature range. It has been proposed that the creep of as-drawn wires occurs by a dislocation mechanism controlled by grain boundary or pipe diffusion (17). The creep-rate exponent n for the ST300-reinforced composites ranged between 5 and 6, which is in agreement with values predicted by simple theories of dislocation climb where n is about 5. It is unlikely that grain boundary diffusion creep or grain boundary sliding controls creep because of the oriented grain structure of the wires.

The relationship between rupture life of the ST300-fiber-reinforced composites and the minimum creep rates is shown in Fig. 26 where a linear inverse relationship was observed of the form:

$$t_R = \frac{C}{\dot{\epsilon}_m}$$

where $C = 0.036$. This relationship has been observed in other metallic materials and is known as the Monkman-Grant relationship (18). This relationship was found to be valid for stainless steel composites reinforced with tungsten-thoria fibers and for nickel-coated and uncoated tungsten-thoria wires (19 and 20). The C value of 0.036 observed in this investigation compares favorably with the value of 0.0207 observed for the

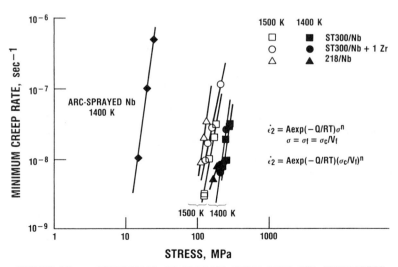

FIGURE 25. - COMPARISON OF MINIMUM CREEP RATE FOR COMPOSITES.

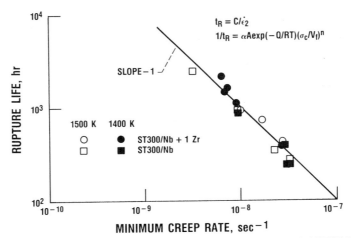

$t_R = C/\dot{\epsilon}_2$

$1/t_R = \alpha A \exp(-Q/RT)(\sigma_c/V_f)^n$

SLOPE-1

RUPTURE LIFE, hr

	1500 K	1400 K	
	O	●	ST300/Nb + 1 Zr
	□	■	ST300/Nb

MINIMUM CREEP RATE, sec^{-1}

FIGURE 26. - RELATIONSHIP BETWEEN RUPTURE LIFE AND MINIMUM CREEP RATE FOR COMPOSITES.

nickel-coated and uncoated wires. The minimum creep-rate expression (previously described for the composites) can be substituted in the Monkman-Grant relationship to yield an expression which equates the composite rupture life with the stress on the composite and with the volume-fraction-fiber content as follows:

$$\frac{1}{t_R} = \left(\frac{1}{C}\right)A \; \exp\left(\frac{-Q}{RT}\right)\left(\frac{\sigma_c}{V_f}\right)^n$$

This expression indicates that at a constant applied stress on the composite, increasing the fiber-volume-fraction content will result in increased rupture life values for the composite.

Advanced space power system components will be required to have service and design lives ranging from 7 to over 10 years. In the interest of determining the potential of the composite materials for such applications, extrapolations for both the 1000- and 100 000-hr (11.4 years) density-corrected creep stress to yield 1-percent strain at 1400 and 1500 K and are compared with similar extrapolations for PWC-11 and Nb-1Zr (Fig. 27). These projections show that the composites are an order of magnitude stronger than Nb-1Zr at 1400 and 1500 K. Compared to PWC-11, the composites are five to six times stronger at 1400 K and three to five times stronger at 1500 K. The strength to density values projected for the composites indicate a potential mass savings that could be realized by use of the composites to replace thicker sections of Nb-1Zr. Alternatively, the potential for increased service temperatures and/or service life of components can be considered.

Future Research

The results to date show that for the SP-100 GES, PWC-11 has attractive creep properties that will extend the capabilities of SP-100 compared to a similar system fabricated from Nb-1Zr. However, additional research is needed in the areas of alloy processing, chemistry control, and heat treatment; establishing uniaxial- and biaxial- (tube) creep data bases; long term aging effects in vacuum and lithium; joining process development; and irradiation testing. Our emphasis on advanced materials for future space power

18

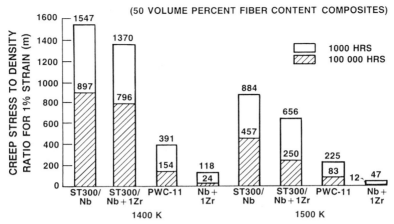

FIGURE 27. - COMPARISON OF PROJECTED 1000 AND 100 000 HR CREEP
STRESS TO DENSITY RATIO FOR 1 PERCENT STRAIN.

systems will continue to focus on the tungsten-reinforced niobium-alloys
composite materials. Follow-on research will explore the effects of angle
plies on creep behavior, fiber-matrix reactions, alternate fibers such as
molybdenum base alloys to reduce composite density, and matrix alloying to
minimize fiber-matrix reaction.

Concluding Remarks

The results to date from our research in support of the Ground Engineering System for SP-100 and our advanced materials technology program for
future space power systems can be highlighted as follows:

1. Based on its demonstrated strength advantage, PWC-11 (Nb-1Zr-0.1C)
has been selected for the SP-100 reference flight system over the weaker
Nb-1Zr alloy.

2. Based on creep rupture and compatibility at 1500 K, tungsten fibers
are potential reinforcements for Nb-base alloys for space power systems.

3. Tungsten-reinforced Nb-1Zr composites provide a ten-fold and a four-
fold creep strength advantage over Nb-1Zr and PWC-11, respectively, at
1400 to 1500 K.

References

1. J.G. Slaby: NASA TM-100795, National Aeronautics and Space
   Administration, Washington, DC, 1988.

2. SP-100 Space Reactor Safety, DOE/NE-0083, Department of Energy,
   Washington, DC, 1987.

3. R.H. Cooper Jr. and E.E. Hoffman, eds.: Refractory Alloy Technology
   for Space Nuclear Power Applications, CONF-8308130, Department of
   Energy, Washington, DC, 1984.

4. R.E. English: NASA TM-89863, National Aeronautics and Space Administration, Washington, DC, 1987.

5. G.R. Dochat: in Proceedings of the Twenty-Second Automotive Technology Development Contractors' Coordination Meeting, SAE P-155, Society of Automotive Engineers, Warrendale, PA, 1984, pp. 209-213.

6. D.R. Hull, D.L. Alger, T.J. Moore, and C.M. Scheuermann: NASA TM-88974, National Aeronautics and Space Administration, Washington, DC, 1987.

7. T.L. Grobstein and R.H. Titran: NASA TM-100848, DOE/NASA/16310-6, National Aeronautics and Space Administration, Washington, DC, 1988.

8. Advanced Materials Program for November and December 1964, PWAC-1018, Pratt & Whitney Aircraft Corp., Middletown, CT, 1965.
9. R.H. Titran, T.J. Moore, and T.L. Grobstein: NASA TM-88842, National Aeronautics and Space Administration, Washington, DC, 1986.

10. T.J. Moore, R.H. Titran, and T.L. Grobstein: NASA TM-88892, DOE/NASA 16310-1, National Aeronautics and Space Administration, Washington, DC, 1986.

11. D.W. Petrasek and R.A. Signorelli: NASA TN D-5139, National Aeronautics and Space Administration, Washington, DC, 1969.

12. D.W. Petrasek: NASA TN D-6881, National Aeronautics and Space Administration, Washington, DC, 1972.

13. L.J. Westfall: NASA TM-86917, National Aeronautics and Space Administration, Washington, DC, 1985.

14. F.G. Arcella: NASA CR-134490, National Aeronautics and Space Administration, Washington, DC, 1974.

15. L.J. Westfall, D.W. Petrasek, D.L. McDanels, and T.L. Grobstein: NASA TM-87248, National Aeronautics and Space Administration, Washington, DC, 1986.

16. D.W. Petrasek and R.H. Titran: NASA TM-100804, DOE/NASA/16310-5, National Aeronautics and Space Administration, Washington, DC, 1988.

17. S.L. Robinson and O.D. Sherby: Acta Met., 1969, vol. 17, pp. 109-125.

18. F.C. Monkman and N.J. Grant: ASTM Proc., 1956, vol. 56, pp. 593-620.

19. R. Warren and L.O.K. Larsson: Matrix Composition and Fiber/Matrix Compatibility in W-Wire Reinforced Composites. Proceedings of the Scandinavian Symposium in Materials Science, University of Lulea, Sweden, 1980.

20. R. Warren and C.H. Andersson: High Temp. High Press., 1982, vol. 14, pp. 41-51.

EFFECT OF MATRIX COMPOSITION ON THE MECHANICAL PROPERTIES OF

TUNGSTEN FIBER REINFORCED NIOBIUM METAL-MATRIX COMPOSITES

T.L. Grobstein

National Aeronautics and Space Administration
Lewis Research Center
MS 49-1
21000 Brookpark Road
Cleveland, OH 44135-3191
216/433-5524

Abstract

The effect of fiber orientation and matrix alloy composition on the creep resistance tungsten fiber reinforced niobium metal-matrix composites was evaluated. When the fibers were oriented ±15° to the stress axis of the composite specimens, the creep resistance decreased ~25% vs composites with axially oriented fibers. Small alloying additions of zirconium and tungsten to the niobium matrix affected the creep resistance of the composites only slightly. In addition, Kirkendall void formation was observed at the fiber/matrix interface; the void distribution differed depending on the fiber orientation.

Refractory Metals: State-of-the-Art 1988
Edited by P. Kumar and R.L. Ammon
The Minerals, Metals & Materials Society, 1989

Introduction

Advanced materials will play a major role in meeting the stringent weight, size, and performance requirements of future space power systems. The requirements for such a system, which may include a service life of greater than 7 years at a temperature in excess of 1350 K in addition to resistance to liquid metal corrosion, dictate the use of refractory metals. The niobium-1 w/o zirconium (Nb-1Zr) alloy has been suggested for use in such space power conversion applications, but does not have sufficient creep strength for the current design times and temperatures. Further, while current designs of space nuclear power systems specify niobium base alloys for reactor, heat pipe, and power conversion components, future applications will need materials with greater high-temperature strength and increased creep resistance to meet the mission requirements (Titran 1988).

The goal of this research is to improve the creep strength/density ratio of the Nb-1Zr alloy without sacrificing desirable physical properties such as corrosion resistance to liquid alkali metals. Fiber reinforcement has been shown to be an effective strengthening method (McDanels et al, 1960), and does not require large alloying additions to the matrix. Therefore, the feasibility of using metal-matrix composites to meet the anticipated increased temperature and creep resistance requirements imposed by advanced space power systems is of interest. The tensile properties of tungsten fiber reinforced niobium metal matrix (W/Nb) composites have been determined by Westfall et al. (1986), and preliminary creep-rupture results were reported by Petrasek and Titran (1988). Both of these studies demonstrated significant improvements in the high-temperature strength/density ratio and the possibility for corresponding mass reductions in high-temperature space power systems.

The current study was designed to measure the creep properties of W/Nb composites and to evaluate the effect of the fiber/matrix reaction on these properties. Two factors which may affect the fiber/matrix interface, and therefore, the creep properties were also evaluated; small matrix alloying additions of tungsten and zirconium, and the orientation of the tungsten fibers.

Materials and Procedures

Tungsten Fibers

When tungsten is drawn into a fine wire, the resulting microstructure may have up to five times the strength of the monolithic material (Yih and Wang 1979). The stability of the fibrous grain structure at elevated temperatures is dependent on the grain-boundary pinning mechanism in the wire. The 100-hour creep rupture strengths of three tungsten-based alloy wires are compared to that for unreinforced (monolithic) Nb-1Zr in Figure 1, and micrographs of their longitudinal sections are in Figure 2.

The 218CS tungsten wire evaluated here is a potassium-doped, commercially available lamp filament. Potassium bubbles pin the grain boundaries, resulting in an increased recrystallization temperature over monolithic (undoped) tungsten (Welsch et al 1979). The ST300 wire is dispersion strengthened with 1.5 weight percent (w/o) thoria (~0.1μm diameter). This wire is slightly stronger than 218CS, and is also commercially available. The W-HfC wire is strengthened by a very fine dispersion (0.05μm diameter) of more thermally stable hafnium carbide particles, and exhibits superior creep properties. Unfortunately, W-HfC wires were not

22

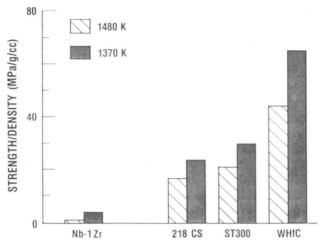

FIGURE 1. - COMPARISON OF THE 100-HR CREEP RUPTURE
STRENGTH-TO-DENSITY RATIO FOR THREE TUNGSTEN-
BASED ALLOY WIRES VERSUS MONOLITHIC Nb-1Zr
ALLOY.

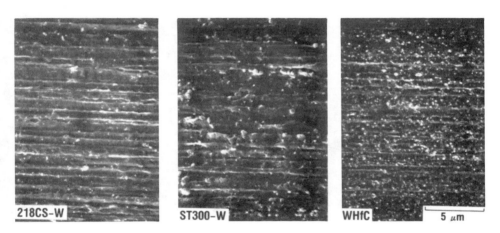

FIGURE 2. - LONGITUDINAL SECTIONS OF AS-DRAWN TUNGSTEN-BASED ALLOY WIRES.

available at the time of this study; therefore, ST300 tungsten wire was
chosen as the best available reinforcement material. The wire had a diameter of 200μm.

23

Matrix Materials

The matrix alloy compositions in weight percent (w/o) and atomic percent (a/o) are listed in Table I. Additions of zirconium (1 and 2 w/o) and tungsten (1, 5, and 9 w/o) were made to niobium to assess their effect on the fiber/matrix reaction. The small addition of zirconium results in a zirconia precipitate which strengthens the alloy and removes oxygen from the grain boundaries of the niobium. Tungsten was added to decrease the concentration gradient between the fiber and matrix, and is a solid solution strengthener in niobium.

Table I. **Matrix Alloy Compositions**

	w/o		a/o	
	Zr	W	Zr	W
Nb				
Nb-1Zr	1.1		1.1	
Nb-2Zr-1W	1.8	1.1	1.8	0.6
Nb-1Zr-5W	0.7	4.7	0.7	2.4
Nb-1Zr-9W	1.3	8.8	1.4	4.6

The matrix materials were obtained in the form of 1.6 mm diameter wire. Specimens of these alloy wires were given a recrystallization anneal and single tensile tests were conducted for each composition at 1366 and 1589 K. The UTS/density ratios for these alloys are given in Figure 3.

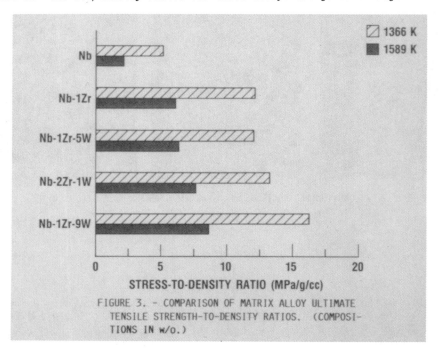

FIGURE 3. - COMPARISON OF MATRIX ALLOY ULTIMATE TENSILE STRENGTH-TO-DENSITY RATIOS. (COMPOSITIONS IN w/o.)

Fabrication

All of the composite panels tested in the program were fabricated using an arc-spray process developed at the NASA Lewis Research Center by Westfall (1985). This process, combined with hot pressing or hot isostatic pressing, produces a fully densified consolidated structure with negligible fiber/matrix interfacial reaction (Figure 4). Composite panels were fabricated with fibers oriented two ways, as shown in Figure 5. The majority of panels had unidirectionally aligned, axially oriented fibers, as well as a limited number with angle-plied (±15°) fibers. Since the fiber content varied from ~35 to 52 volume percent (v/o), applied creep stresses were normalized to 50 v/o for comparison purposes.

FIGURE 4. - AS-FABRICATED MICROSTRUCTURE OF A W/Nb COMPOSITE CONTAINING ~50 v/o FIBER.

Testing

Specimens having a reduced gauge section 25.4 mm long and 6.35 mm wide were machined from the composite panels. Creep-rupture tests were conducted in a vacuum of 7×10^{-5} Pa. Creep strains were measured optically via a cathetometer clamped to the furnace chamber frame sighting on Knoop hardness impressions placed 2.54 cm apart in the reduced section. The creep apparatus is fully described elsewhere by Titran and Hall (1965). The precision of creep strain measurements is estimated to be ~0.02% for the gauge length used. The strain on loading was measured and was incorporated in the reported total creep strain. Figure 6 depicts a fractured creep specimen.

25

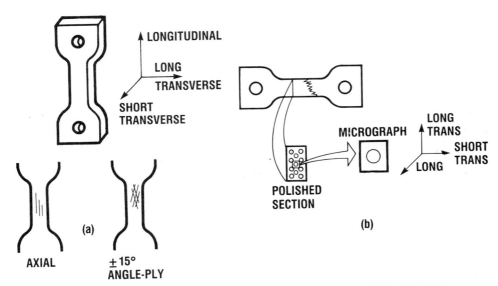

FIGURE 5. - SCHEMATIC OF THE FIBER ORIENTATIONS IN W/Nb CREEP SPECIMENS.

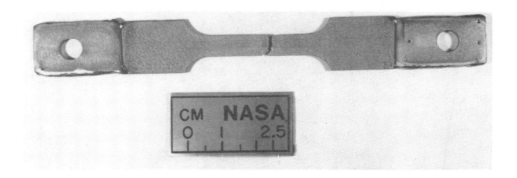

FIGURE 6. - TYPICAL FAILED W/Nb COMPOSITES CREEP SPECIMEN. TUNGSTEN TABS (0.5 MM THICK) WERE ELECTRON BEAM WELDED ONTO THE ENDS OF THE SAMPLES TO PREVENT SHEAR PULLOUT OF THE PINHOLE AREA.

Results and Discussion

Composite Creep Properties

Typical fracture surfaces of creep-tested specimens are shown in Figure 7. In these micrographs, the matrix appears to undergone ductile failure, whereas the fibers failed in a brittle manner. In fact, the matrix in these composites undergoes plastic deformation early in the lifetime of the composite, and functions only as a protective coating during the remainder of the creep life. When the fibers fail, the entire composite fails. The creep behavior of the composite will therefore resemble that of the tungsten fibers. This behavior is demonstrated in Figure 8, a plot of a typical creep curve for a W/Nb composite sample.

A plot of the time to rupture vs the normalized applied stress (Figure 9) compares the composites at 1400 K and 1500 K, and the minimum creep rate taken from the creep curves is plotted in Figure 10. Various aspects of these data will be discussed in the following sections.

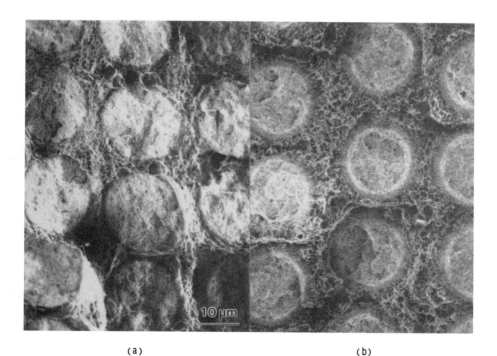

(a)   (b)

FIGURE 7. - FRACTURE SURFACES OF ST300/Nb-1Zr COMPOSITES CREEP TESTED AT 1500 K UNDER AN APPLIED STRESS OF (a) 228 MPa FOR 425 HR AND (b) 80 MPa FOR 5353 HR. SEM MICROGRAPHS REVEAL THE DIFFERENT FRACTURE MODES OF THE FIBER AND MATRIX.

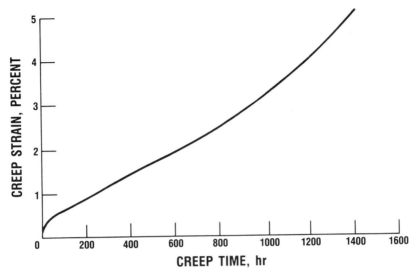

FIGURE 8. - TYPICAL CREEP CURVE FOR A ST300/Nb-1Zr COMPOSITE
TESTED AT 1400 K UNDER A NORMALIZED APPLIED STRESS OF 238
MPA. THE CURVE EXHIBITS PRIMARY, STEADY STATE, AND TERTIARY
CREEP REGIMES.

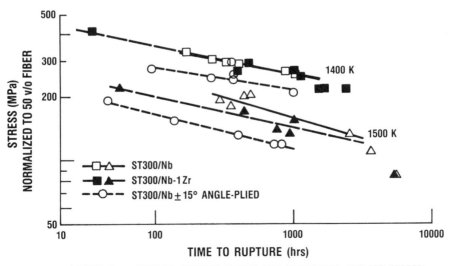

FIGURE 9. - TIME-TO-RUPTURE VERSUS THE NORMALIZED APPLIED STRESS
(50 v/o) FOR W/Nb COMPOSITES CREEP-TESTED AT 1400 AND 1500 K.

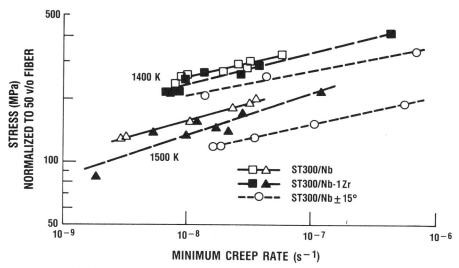

FIGURE 10. - MINIMUM CREEP RATE VERSUS THE NORMALIZED APPLIED STRESS (50 v/o FOR W/Nb COMPOSITES CREEP-TESTED AT 1400 AND 1500 K.

Effect of Interdiffusion

Since the phase diagram for tungsten and niobium is a continuous solid solution, elevated temperature exposure will result in interdiffusion between the fiber and matrix. This interdiffusion can effectively reduce the volume fraction of fiber in the composite, and, as it progresses over time, will degrade the composite's mechanical properties. We must be able to predict this degradation if this composite material is to be used in real systems. If the degradation is substantial, we must look at ways to stop or slow the interdiffusion.

It is apparent in the fracture micrographs (Figure 7) that interdiffusion occurs between the fiber and matrix upon elevated temperature exposure. Microprobe traces of the fiber/matrix interface indicate that the majority of the diffusion is into the fiber (Figure 11). When the polished sections are examined (Figure 12), a recrystallized zone is evident within the original outer diameter of the fiber. These facts suggest that niobium diffuses into the fiber and lowers its recrystallization temperature, resulting in a recrystallized and tungsten-depleted area within the original fiber diameter.

In Figures 9 and 10, a negative deviation from the established slope of each of the lines is apparent for specimens subjected to relatively low applied stresses. Since these specimens were tested for correspondingly long times, the negative deviation is attributed to the growth of the fiber/matrix reaction zone. This recrystallized area of the fiber no longer has the high-strength fibrous grain structure, and no longer contributes to the strength of the composite in the capacity it did at the start of the test.

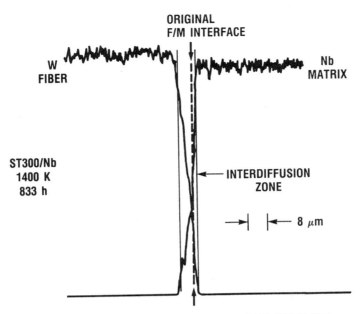

FIGURE 11. - TYPICAL MICROPROBE TRACE ACROSS THE
FIBER/MATRIX INTERFACE IN A ST300/Nb COMPOSITE EX-
POSED TO 1400 K IN VACUUM FOR 833 HR UNDER A NORM-
ALIZED APPLIED STRESS OF 238 MPA. THE DOTTED LINES
INDICATE THE EXTENT OF INTERDIFFUSION TAKING THE
LIMITATIONS OF THE INSTRUMENTATION INTO ACCOUNT.
THE POSITION OF THE ORIGINAL FIBER/MATRIX INTERFACE
IMPLIES THAT THE MAJORITY OF THE DIFFUSION OCCURS
FROM THE MATRIX INTO THE FIBER.

This effect of interdiffusion is further demonstrated if the creep
properties of the composites are compared to that of the fiber itself. In
Figure 13, the stress-rupture properties of ST300 fibers have been extrapolated using the Monkman-Grant (1956) relationship (which states that the
time-to-rupture and the minimum creep rate are inversely proportional) to
plot the theoretical time-to-1%-strain at 1500K for a bundle of fibers
assuming no surrounding matrix. On the same graph, the time-to-1%-strain
for composites with several matrix compositions is plotted. The choice of
time-to-1%-strain as a creep parameter is based on current design requirements space nuclear power systems.

Since the fibers contribute almost all of the composite strength, it
is expected that the values for the composite and the fibers would differ
only slightly. It is apparent from Figure 13 that these values would be
comparable if the test time were extrapolated back to zero. As time prog-
ressed, however, the composite properties diverged negatively from the
theoretical values calculated for the fibers. This can be attributed to
the fact that the growing reaction zone removes an increasing amount of the
high-strength fiber microstructure; therefore, a smaller volume fraction of
fiber remains to contribute to the strength of the composite.

30
.

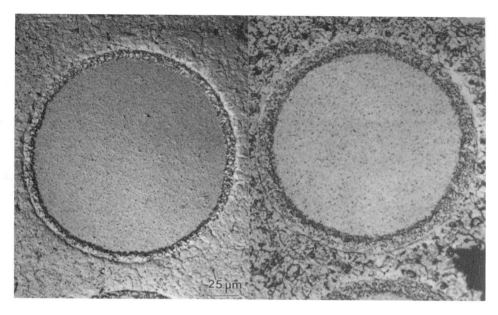

(a)   (b)

FIGURE 12. - TRANSVERSE SECTIONS OF ST300/Nb-1Zr COMPOSITES CREEP-TESTED AT 1500 K
UNDER AN APPLIED STRESS OF (a) 131 MPa FOR 2549 HR AND (b) 80 MPa FOR 5353 HR.
METALLOGRAPHIC PREPARATION ILLUSTRATES THE REACTION ZONE BETWEEN THE FIBER AND
MATRIX.

Inhibiting Interdiffusion

One way to impede interdiffusion in the composite is to create a diffusion barrier between the fiber and matrix. This technique was investigated by Veltri et al. (1975), who ion-plated thermally stable coatings
(such as Al_2O_3, Y_2O_3, and HfO_2) onto tungsten wires. Unfortunately, when
incorporated into a composite as coated fibers, small cracks in the coatings occurred due to the differential thermal expansion properties between
the matrix, coating, and fiber. It was discovered that these small cracks
in the coating allowed very fast diffusion of the matrix throughout the
fiber/coating interface, leading to fracture of the fiber.

The feasibility of ion-implanted diffusion barriers has also
been explored. Welsch et al (1987) investigated the concept that a layer
of atoms on the fiber surface with considerably larger atomic radii than
either the fiber or matrix material may impede interdiffusion. Although
this technique shows promise (Kopp et al, 1988), it is far from being
applied.

A different method to suppressing the interdiffusion would be to make
alloying additions to the matrix. (Alloying the fiber is impractical due
to the complex chemistry which maintains the high-strength structure of the
wire.) The effect of zirconium in the matrix on the interdiffusion kinetics is of interest, given that Tuchinsky (1979) predicted that alloying
niobium with zirconium would decrease its rate of diffusion into tungsten.

31

Also, it is known that alloying niobium with tungsten decreases the niobium self-diffusion rate (Lyubimov et al 1967; Lundy and Pawel 1969; Mundy et al 1986). It is possible, therefore, that alloying the matrix material with small amounts of tungsten will decrease interdiffusion rates in the composite.

The three lines plotted for the composite properties in Figure 13 represent composites with 0, 1 and 2 w/o zirconium additions to their niobium-base matrices. A small number of creep data points for samples with 5 and 9 w/o tungsten are also plotted. For reasons outlined later, the addition of 1 w/o tungsten was regarded as negligible, and the decrease in strength of the Nb-2Zr-1W alloy was mainly attributed to the zirconium addition.

The data in Figure 13 reveals an increasing negative deviation from the ideal fiber strength with increasing zirconium matrix content, the converse of the tensile results shown in Figure 3, where an increasing zirconium content strengthened the matrix alloys. It is possible that the addition of zirconium enhanced the fiber/matrix reaction, resulting in a decrease in the volume fraction of fiber in the composite, and reducing the strength of the composite at long times. Also, there is little difference in the creep strength between Nb-1Zr with and without 5 w/o tungsten, whereas the addition of 9 w/o tungsten appears to increase the creep resistance of the composite by 10% at a time-to-1%-strain of 100 hr, and up to 24% at a time-to-1%-strain of 3000 hr. This may be ascribed to the higher strength of the matrix (as shown in Figure 3), to a decrease in the rate of diffusion between the tungsten fiber and the niobium alloy matrix, or a combination of the two.

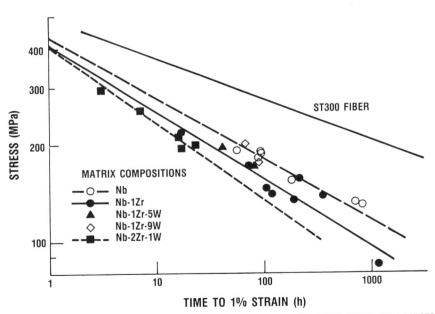

FIGURE 13. - COMPARISON OF THE CREEP STRENGTH OF ST300 FIBER REINFORCED COMPOSITES VERSUS THAT FOR A BUNDLE OF FIBERS WITH AN EQUIVALENT CROSS-SECTIONAL AREA OF FIBER AT 1500 K.

These results indicate that at least 9 w/o tungsten must be added to see any effect of the tungsten addition; however, it is important to note that an addition of 9 w/o tungsten to the matrix increases the density of a 50 v/o fiber composite by approximately 3%. Hence, the increase in real strength of the composite correlates to a somewhat lesser increase in its specific strength. While the addition of tungsten may impart enough additional creep resistance to the composite to make it worthwhile, longer term tests will be necessary in order to confirm and extrapolate this data.

Effect of Fiber Orientation

Maximum strengthening in a continuous-fiber reinforced composite is achieved by orienting all of the fibers in the composite in a single direction, and then orienting the composite so the applied stress is along the fiber axis. However, any real application will usually be a biaxial or triaxial stress state, and therefore require at least some strength along more than one axis. Angle-plying the fibers is an obvious method to increase the transverse properties of the composite. However, the effect of changing the fiber orientation on the mechanical properties must be understood and quantified before the composite can be used in real applications.

In the W/Nb system, the strength of composites with fibers angle-plied ±15° is decreased about 25% compared to that for composites with uni-directional, axially aligned fibers. This is apparent when comparing the time to rupture (Figure 9) and the minimum creep rate (Figure 10).

In addition to recrystallization of the interdiffusion zone, the formation of Kirkendall voids occurs at the fiber/matrix interface. This phenomenon was first observed in the zinc/alpha brass system by Smigelskas and Kirkendall (1947), and is well-documented in the W/Nb system (Leber and Hehemann 1966; Arcella 1974). These workers found that niobium diffuses into tungsten faster than tungsten diffuses into niobium. In the composite, the net mass transfer of niobium into the fiber causes a ring of very small voids to form outside of the original fiber diameter. Over time, these microvoids accumulate, and the microstructure shown in Figure 14a results. These voids at the fiber/matrix interface may affect the composite properties given that the integrity of the interface area in any ductile metal-matrix composite is essential for load transfer from the matrix to the fiber. Although it is apparent from the creep results that these voids do not grossly affect the axial properties of the composites, it is probable that they will have a detrimental effect on the transverse strength.

When the fibers are angle-plied ±15°, these voids form only on two sides of the fiber. This is illustrated in Figure 14b, where the long transverse direction of the composite panel is in the vertical direction in the micrograph. When the fibers are angle-plied, compressive stresses are induced on the fiber due to the forces which are attempting to pull the fibers parallel to the strain axis. These compressive stresses inhibit the formation of Kirkendall voids in some areas around the diameter of the fiber. Voids form in the remaining unstressed areas just as they do in the unidirectionally reinforced composites. Angle-plying of the fibers can therefore augment the transverse properties of the composite in two ways. First, some of the applied stress in the transverse direction is transferred to the longitudinal direction of the fibers. Second, the induced compressive stresses appear to improve the fiber/matrix bond, possibly increasing load transfer across the interface.

33

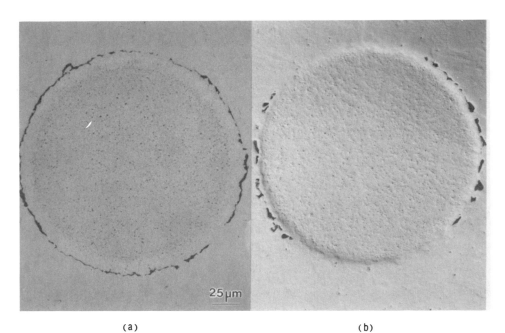

(a)   (b)

ST300/Nb   ST300/Nb ($\pm15^0$ ANGLE-PLIED)
1500 K/131 MPA   1500 K/115 MPA
2485 HR   715 HR

FIGURE 14. - TRANSVERSE SAMPLES IN THE AS-POLISHED CONDITION REVEAL THE PRESENCE OF
KIRKENDALL VOIDS. THE MICROGRAPH IN (a) SHOWS A SINGLE FIBER IN A COMPOSITE PANEL
WITH UNIDIRECTIONAL AXIAL REINFORCEMENT. THE MICROSTRUCTURE EXHIBITS A UNIFORM
DISTRIBUTION OF VOIDS JUST OUTSIDE OF THE ORIGINAL FIBER DIAMETER. THE FIBERS IN
THE PANEL IN (b) WERE ANGLE-PLIED $\pm15^0$ TO THE LONGITUDINAL AXIS, CAUSING COMPRES-
SIVE STRESSES WHICH SUPPRESSED THE FORMATION OF VOIDS IN SOME AREAS AROUND THE
FIBER.

Composite vs Monolithic Material Properties

Figure 15 compares the time-to-1%-strain of unidirectionally reinforced composites to angle-plied composites, monolithic Nb-1Zr, and monolithic PWC-11 alloy (Nb-1Zr-0.06C) at 1500 K (Titran et al, 1988).
Composites with angle-plied fibers ($\pm15°$) are ~25% weaker than
unidirectionally reinforced panels, but exhibit approximately the same
stress exponent, indicating that the same creep mechanisms are at work.
All of the composite materials are approximately an order of magnitude
stronger than monolithic Nb-1Zr, and are ~4 times stronger than the PWC-11
alloy at this temperature.

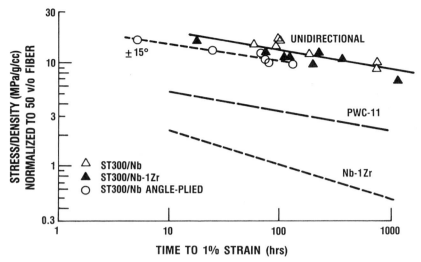

FIGURE 15. - COMPARISON OF THE TIME-TO-1%-STRAIN FOR ST300
FIBER-REINFORCED COMPOSITES AND THE MONOLITHIC Nb ALLOYS
Nb-1Zr AND PWC-11 (Nb-1Zr-0.06C) (AFTER TITRAN, ET AL.,
1988). ALL DATA IS PLOTTED ON A STRESS-TO-DENSITY BASIS,
AND THE COMPOSITE DATA WAS NORMALIZED TO 50 v/o FIBER.

SUMMARY

Reinforcement with ST300 fibers increases the creep resistance of niobium alloys significantly. The fiber/matrix reaction, however, tends to
degrade the composite creep-rupture properties at long times. This
degradation is mainly due to recrystallization of a reaction zone in the
fiber, but it may also be due to related Kirkendall void formation which
can affect load transfer. Although angle-plying the reinforcing fibers
±15° decreases the axial composite creep-rupture properties by ~25%, both
the axially reinforced and angle-plied composite materials still maintain a
clear advantage over conventional monolithic niobium alloys.

Matrix alloying additions of 9 w/o tungsten appear to increase the
creep resistance of W/Nb composites, although this increase is partially
offset by an increase in the density of the composite. There is no indication of an effect of smaller additions of tungsten to the niobium matrix on
the composite properties, but small additions of zirconium apparently cause
a slight increase in the fiber/matrix interdiffusion rates, thereby lowering the creep resistance of the composite. Before these results can be
extrapolated for long-term creep life, they must be confirmed by actual
creep testing of these materials for longer times.

Acknowledgments

Discussions with RH Titran, DW Petrasek, and G Welsch were indispens-
able to the completion of this study. Expert metallography by TA Leonhardt
is also gratefully acknowledged. This work was performed by the NASA Lewis
Research Center for NASA and the Strategic Defense Initiative Office under
an interagency agreement entered into in October 1985.

References

F.G. Arcella, **"Interdiffusion Behavior of Tungsten or Rhenium and Group V and VI Elements and Alloys of the Periodic Table,"** Part 1 (Report NASA CR-134490, Westinghouse Astronuclear Laboratory, Pittsburgh, PA, 1974).

R.F. Hehemann and S. Leber, **"Chemical Diffusion in the Columbium-Tungsten System,"** Trans AIME 236 (1966) 1040-1044.

M.W. Kopp and J.K. Tien, **"A Preliminary Study of Ion-Implanted Diffusion Barriers in High Temperature Metal Matrix Composites,"** Scripta Met 22 (1988) 1527-1530.

M.W. Kopp, J.K. Tien, and D.W. Petrasek, **"Reaction Kinetics between Fiber and Matrix Components in Metal Matrix Composites,"** Superalloys 1988, ed. S. Reichman, D.N. Duhl, G. Maurer, S. Antolovich, and C. Lund (Warrendale, PA: TMS, 1989) 193-201.

T.S. Lundy and R.E. Pawel, **"Effects of Short-Circuiting Paths on Diffusion Coefficient Measurements,"** Trans AIME 245 (1969) 283-286.

V.D. Lyubimov, P.V. Gel'd, G.P. Shveykin, and Yu.A. Sutina, **"Self-Diffusion of Niobium in its Alloys with Tungsten,"** Isvestiya AN SSSR, Metally 2 (1967) 84-87. Translated in Russian Metallurgy (1967) 40-43.

D.L. McDanels, R.W. Jech, and J.W. Weeton, **"Metals Reinforced with Fibers,"** Metal Progress 78 (1960) 118-121.

F.C. Monkman and N.J. Grant, **"An Empirical Relationship between Rupture Life and Minimum Creep Rate in Creep-Rupture Tests,"** ASTM Proceedings 56 (1956) 593-605.

J.N. Mundy, S.T. Ockers, and L.C. Smedskjaer, **"Dehancement of Impurity and Self-Diffusion in Niobium by Tungsten Additions,"** Phys Rev B 33 (1986) 847-853.

D.W. Petrasek and R.H. Titran, **"Creep Behavior of Tungsten/Niobium and Tungsten/Niobium-1%Zirconium Composites"** (Report DOE/NASA/16310-5, NASA TM-100804, NASA Lewis Research Center, Cleveland, OH, 1988).

A.D. Smigelskas and E.O. Kirkendall, **"Zinc Diffusion in Alpha Brass",** Trans AIME 171 (1947) 130-142.

R.H. Titran and R.W. Hall, **"High-Temperature Creep Behavior of a Columbium Alloy, FS-85",** (Report NASA TN D-2885, NASA Lewis Research Center, Cleveland, OH, 1965).

R.H. Titran, J.R. Stephens, and D.W. Petrasek, **"Refractory Metal Alloys and Composites for Space Nuclear Power Systems"** (Report DOE/NASA/16310-8, NASA TM-101364, NASA Lewis Research Center, Cleveland, OH, 1988).

L.I. Tuchinsky, **"Thermodynamic Method of Calculating the Effect of Alloying Additives on Interphase Interaction in Composite Materials,"** Fizika i Khimia Obrabotka Materialov 1 (1979) 121-126. Translated in Report NASA TM-88391, 1986.

R.D. Veltri, E.L. Paradis, and F.C. Douglas, **"Investigation to Develop a Method to Apply Diffusion Barriers to High Strength Fibers"** (Report NASA CR-134719, United Aircraft Research Laboratory, East Hartford, CT, 1975)

G. Welsch, B.J. Young, and R.F. Hehemann, **"Recovery and Recrystallization of Doped Tungsten"**, <u>Strength of Metals and Alloys V</u>, ed. P. Haasen, V. Gerold, and G. Kostorz (New York, NY: Pergamon Press, 1979), 13-18.

G. Welsch, K.T. Kim, and J.J. Wang, **"Potassium-Implanted Tungsten Fibers in Nickel-Base Superalloy for Study of Reaction Barrier"** (Final Report for NASA NAG 3-67, Case Western Reserve University, Cleveland, OH, 1987).

L.J. Westfall, **"Tungsten Fiber Reinforced Superalloy Composites Monolayer Fabrication by an Arc-Spray Process"** (Report NASA TM-86917, NASA Lewis Research Center, Cleveland, OH, 1985).

L.J. Westfall, D.L. McDanels, D.W. Petrasek, and T.L. Grobstein, **"Preliminary Feasibility Studies of Tungsten/Niobium Composites for Advanced Space Power Systems Applications"** (Report NASA TM-87248, NASA Lewis Research Center, Cleveland, OH, 1986).

S.W.H. Yih and C.T. Wang, <u>Tungsten</u> (New York, NY: Plenum Press, 1979) 270-299.

TENSILE AND CREEP-RUPTURE BEHAVIOR OF

P/M PROCESSED NB-BASE ALLOY, WC-3009

MOHAN G. HEBSUR
SVERDRUP TECHNOLOGY INC.,
MIDDLEBURG HTS, OH 44130

ROBERT H. TITRAN
NASA LEWIS RESEARCH CENTER
CLEVELAND, OH 44135

ABSTRACT

Due to its high strength at temperatures up to 1600 K, fabrication of niobium base alloy WC-3009 (Nb-30Hf-9W) by traditional methods is difficult. Powder metallurgy (P/M) processing offers an attractive fabrication alternative for this high strength alloy. Spherical powders of WC-3009 produced by electron beam atomizing (EBA) process were successfully consolidated into one inch diameter rod by vacuum hot pressing and swaging techniques.

Tensile strengths of the fully dense P/M material at 300-1590 K were similar to the arc-melted material. However, the P/M material showed somewhat less tensile elongations than the arc-melted material. Creep-rupture tests in vacuum indicated that WC-3009 exhibits a Class I solid solution (glide controlled) creep behavior in the 1480-1590 K temperature range and stress range of 14 to 70 MPa. The creep behavior was correlated with temperature and stress using a power law relationship. The calculated stress exponent n, was about 3.2 and the apparent activation energy, Q , was about 270 kJ/mol. The large creep ductility exhibited by WC-3009 was attributed to its high strain rate sensitivity.

Refractory Metals: State-of-the-Art 1988
Edited by P. Kumar and R.L. Ammon
The Minerals, Metals & Materials Society, 1989

INTRODUCTION

Niobium (Nb) base alloys are attractive for advanced aerospace propulsion applications because of their favorable combination of low density, high melting point and elevated temperature mechanical properties. However, none of the currently available commercial Nb alloys have the desired combination of strength, ductility and oxidation resistance required for these applications. High strength alloys lack ductility, oxidation resistance and are difficult to fabricate whereas ductile alloys lack sufficient strength at high temperatures. Oxidation resistant alloys such as niobium aluminides are hard and brittle. If significant improvements in strength, ductility and oxidation resistance are to be achieved for structural alloys for advanced aerospace propulsion components, it is most probable that these improvements will be provided by innovation in processing and production technology. An understanding of the structure and composition of the protective oxide scales on niobium alloys combined with unique processing techniques such as rapid solidification and mechanical alloying, the desired materials can be produced.

Thus the NASA-Lewis Research Center has undertaken a broad program to develop niobium base alloys with improved oxidation resistance, ductility and strength for advanced aerospace applications. In this overall program, a commercially available high strength Nb-alloy WC-3009 (Nb-30Hf-9W) in a powder form and the most oxidation resistant niobium aluminide intermetallic compound developed (1) will be mechanically alloyed in a high energy attritor mill. The mechanical alloying hopefully will cause the brittle intermetallics to fragment and embed in the surface of the more ductile WC-3009 alloy. It is felt that this type of microstructure of powder particle may help in selective oxidation of Al from the well dispersed intermetallic phase to form a more protective oxide scale. However, the addition of a brittle intermetallic to ductile alloy may degrade some of its mechanical properties. Therefore, the aim of the present investigation was first to identify the powder processing parameters and then to evaluate the tensile and creep-rupture behavior of P/M processed WC-3009.

EXPERIMENTAL PROCEDURE

Spherical powders of WC-3009, procured commercially from Teledyne Wah-Chang Albany (TWCA), were produced by electron beam melting followed by centrifugal atomization in a laboratory size electron beam atomizer. The chemical analysis of as-received powder is given in Table I.

Table I
Chemical Composition Of As-Received WC-3009 Powder
(In Wt%)

Nb	Hf	W	Zr	Ta	O	N	C
56.0	33.2	9.6	1.02	0.5	0.020	0.0015	0.0085

The spherical powders were canned in niobium tubes, 38 mm dia and 500 mm long, followed by vacuum hot pressing at 1590 K under a pressure of 200 MPa for 8 hours. The bars were then hot and warm swaged to 12.5 mm dia rod. The fully consolidated rods were vacuum annealed at 1590 K for one hour. With an intention of investigating the influence of temperature and pressure on the consolidation of the powders, two trials of hot pressing were carried out at 1590 and 1810 K under a pressure of 150 MPa for 8 and 3 hours respectively.

ASTM standard tensile and creep specimens of 25.4 mm gage and 6 mm dia round were machined from the heat treated rod. Sheet specimens of 25.4 mm gage were also machined from recrystallized sheets of WC-3009 obtained from TWCA. These sheets were produced by conventional techniques of arc melting and rolling (hereafter referred to as arc-melted). Tensile tests were carried out according to ASTM E-8 procedures. Elevated temperature tensile tests were performed under high vacuum. Constant load creep rupture tests were conducted in a vacuum of 10^{-4} MPa in the temperature range 1480 to 1590 K and stress range 14 to 70 MPa using apparatus described in detail elsewhere (2). The specimens were heated to test temperature by radiation from a concentric tantalum sheet heater positioned within a water cooled chamber. The temperature was monitored with Pt/Pt-Rh thermocouples attached the reduced gage section of the specimens and to a digital temperature display. Creep strains were measured optically using a cathotometer to sight on fiducial marks initially placed on the gage length of the specimen. The precision of creep strain measurements is estimated to be ± 0.02% for the gage length used. Creep strains were also monitored on a dial micrometer attached to the pull rods. This was particularly useful in monitoring large elongations for most tests.

RESULTS

The particle size analysis of the as-received powder given in Table II suggests that most of the powder particles had an average size of about 63 μm.

Table II			
Particle Size Distribution Of As-Received WC-3009 Powder.			
Size (μm)	88	63	35	25
Percent	4	60	15	11

Figure 1 shows the microstructure of as-received powder particles. Most of the powder particles were spherical and showed evidence of coring in the dendritic microstructure. This suggests that the particles had undergone high cooling rates of 10^{3-4} K/sec in the interdendritic areas.

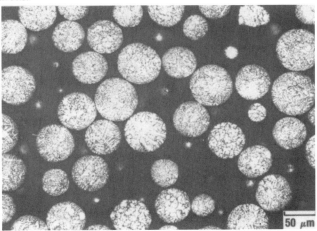

50 μm

Figure 1: Microstructure of as−received powders of niobium alloy,WC−3009.

The consolidation of powders by hot pressing at 1590 K under a pressure of 200 MPa followed by swaging produced a fully dense rod. However, it was not possible to get a full densification when the hot pressing was carried out at 1590 K using a pressure of 150 MPa. Increasing the hot pressing temperature to 1810 K resulted in a severe reaction with the niobium pressing cans. The microstructure of the fully consolidated and annealed rod consisted of fine grains (Figure 2). The second phase particles present in Figure 2 were identified as hafnium oxide, formed due to the high oxygen content in the powder (Table I).

Figure 2: Optical micrograph of the fully consolidated and
and 1 hour 1590 K vacuum annealed WC−3009
P/M alloy.

Figure 3 shows the results of tensile tests on both the arc-melted and P/M processed WC-3009. The values of strength and ductility shown in Figure 3 were the average of duplicate specimens tested under the identical conditions of temperature and strain rate. The range of tensile strengths levels observed in the P/M material at all test temperatures were within the normal range for arc-melted material. However, tensile elongations were affected by processing methods.The P/M material exhibited less tensile ductility than the arc-melted material at all test temperatures. Both arc-melted and P/M material showed a sharp rise in tensile elongation with temperature beyond 1480 K probably due to dynamic recrystallization. Wojcik (3) has observed a decrease in ductility in both forms of this alloy at 1380-1480 K. However this was not observed in the present investigation.

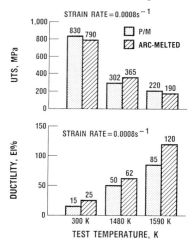

Figure 3: Comparison of tensile properties of arc−melted
and P/M processed niobium alloy, WC−3009.

42

A typical creep curve of P/M processed WC-3009 tested at 1480 K under an initial stress of 55 MPa is shown in Figure 4. This creep curve shows that the material exhibited very little primary creep, a relatively short steady state region and a large amount of tertiary creep which resulted in high strains to fracture. This type of creep behavior was exhibited over the entire temperature and stress levels tested for both the arc-melted and P/M processed WC-3009 alloy.

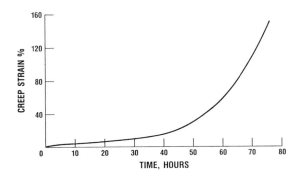

Figure 4: A typical creep curve of P/M processed niobium alloy, WC-3009 tested at 1480 K and 55 MPa.

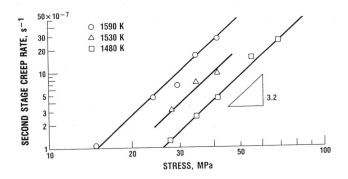

Figure 5: Steady state creep rate as a function of applied stress and temperature for niobium alloy, WC-3009.

The creep properties were characterized in terms of the stress and temperature dependence of the steady-state creep rate, $\dot{\epsilon}$, and the rupture life, t_f. The steady state creep rate can be described (4) by the following phenomenological equation:

$$\dot{\epsilon} = A\sigma^n \exp\left(-\frac{Q}{RT}\right)$$ [1]

where A is a material constant, n is the stress exponent and Q is the apparent activation energy for creep, R is the universal gas constant, and T is the absolute temperature in K. This power law relation between steady-state creep rate and applied stress is illustrated in Figure 5 for P/M processed WC-3009 tested at 1480, 1530 and 1590 K. The values of n and Q were estimated by multiple regression analysis using all the creep data obtained at the three temperatures. The coefficients, n and Q, of the regression equation were found to be equal to 3.15 ± 0.21 and 270 ± 40 kJ/mol respectively.

In addition, the stress and temperature dependence of the creep rupture lives can be described accurately (5) by the following equation:

$$t_f = A\sigma^P \exp\left(-\frac{Q}{RT}\right)$$ [2]

which is of the same form as eqn. [1]. The linear behavior of the log t_f vs log σ data is illustrated in Figure 6 in the temperature range 1480 and 1590 K.The coefficients, p and Q, determined by multiple regression were found to be equal to 3.9 ± 0.22 and 370 ± 32 kJ/mol, respectively.

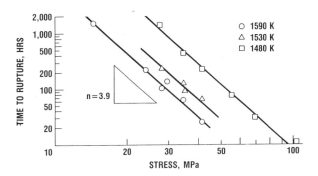

Figure 6: Time to rupture as a function of applied stress and temperature for niobium alloy, WC-3009.

The activation energies were also determined with the corrections for the temperature dependence of the elastic modulus, E. In this case, the temperature dependence of E for pure niobium from Reference 6 were used. Using the modulus correction in eqn. [1] and in eqn. [2] the apparent activation energies of 240 ± 30 and 338 ± 32 kJ/mol respectively were obtained. Thus the modulus correction did not significantly influence the values of activation energies determined for both creep rate and rupture life.

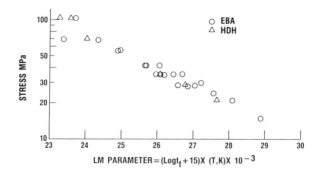

Figure 7: Larson–Miller plot showing the agreement between the atomized and hydride–dehydride powder processed WC–3009 alloy.

A summary of stress-rupture data of WC-3009 is shown as a Larson-Miller plot in Figure 7. The stress-rupture data on hydride-dehydride (HDH) powder processed WC-3009 reported in Reference 3 is also included in Figure 7. As with most niobium alloy data, a constant of 15 was used in the Larson-Miller equation. As shown in this figure, there was a good agreement between the rupture lives of the atomized and hydride-dehydride powder processed material. A fairly wide scatter band evident in Figure 7 may be due to scatter in tertiary creep which dominated most of the tests. It should also be recognized that considerable differences in total elongations were observed as evident in Figure 8 and no correlation between elongations and applied stress could be made. Figure 8 also indicates a larger amount of second phase particles present in a specimen tested at 14 MPa (which failed after 1465 hours) than in a specimen tested at 35 MPa (which failed after 64 hours) suggesting that these particles may be responsible for decreased ductility. The chemical analysis of the specimen tested at 14 MPa indicated an increase in oxygen content by about 60 ppm. Microprobe results confirmed the second phase particles to be hafnium oxide.

DISCUSSION

Although refractory metal powders have long been used by consolidating into sintered electrodes for subsequent melting into ingots for traditional metallurgical processing, application of powder metallurgy techniques as an alternate to the traditional techniques to produce a near-net-shape component is a recent one. Results of earlier work (3,7) as well as the present work suggest that P/M approach is a viable one for attaining properties equivalent to wrought material and is capable of directly producing near net shape components of the high strength niobium alloys which are otherwise difficult fabricate, thus, significantly saving the strategic refractory materials. The WC-3009 alloy is an excellent candidate since this alloy is very brittle in cast form, with a typical ductility < 1%. However, at least 15% tensile elongation can be achieved in a material consolidated from rapidly solidified powder without any subsequent hot or cold working (3). With the current

 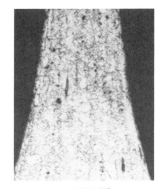

1590 K/35 MPa   **1590 K/14 MPa**

Figure 8: Comparison of cross-sections of fractured
specimens tested at 35 and 14 MPa at 1590 K.

interest in niobium intermetallics for advanced aerospace applications due to their low density and good oxidation resistance, it is believed that only P/M methods will be capable of producing improved compositions.

The dendritic microstructure of the powder particles shown in Figure 1 is typical of PREP (Plasma Rotating Electrode Process) and EBA (Electron Beam Atomized) powders which undergo intermediate cooling rates of 10^{3-4} K/sec. For comparison, melt-spun products which undergo very high cooling rates ($\approx 10^6$ K/sec) and hence exhibit a more homogeneous and refined microstructure, resulted in improved strength and ductility (8). Wojcik (3) has shown that the WC-3009 powder produced by hydride-dehydride process was angular in shape and had the same structure observed in the as-cast ingot.

The tensile test data shown in Figure 3 indicate that P/M processed material results in strengths equivalent to wrought material at room and elevated temperatures. Similar results have been observed for niobium alloy C-103 (7). However, P/M alloy WC-3009 exhibited a decrease in ductility at all temperatures as compared to the arc-melted material. It is known that the distribution of oxides on the grain boundaries can drastically reduce tensile elongations (3). The P/M processed WC-3009 had higher amounts of oxygen (TableI) in the form of finely distributed hafnium oxide (Figure 2) which may be responsible for the somewhat lower ductility.

The shape of the creep curve shown in Figure 4 in which there is a very large region in which the strain rate is continuously increasing with the strain, is a typical of Class I solid solution strengthened alloy as opposed to pure metals and Class II or climb controlled alloys (9). This type of creep curve has been documented for another niobium alloy, C-103 (7). In contrast with pure metals and other alloys, Class I solid solution strengthened alloys exhibit subgrain formation only at very large strains and are not controlled by stacking fault energy (9,10).

The creep behavior of this alloy is similar to that of C-103 in which the values of n of 3.4 and Q of 316 kJ/mol, a value smaller than the activation energy for self-diffusion are observed (7). The creep behavior of Class I solid solution alloys is generally considered to be controlled by solute viscous drag on gliding dislocations (9-11). Therefore, the activation energy for creep might be expected to be related to the diffusion of either tungsten or hafnium in niobium. The diffusion of tungsten in niobium for up to 1473 K is only 38 kJ/mol (12) which makes it unlikely that this element controls creep. Unfortunately an activation energy value for the diffusion of hafnium in niobium is not found in literature.

The stress σ and strain rate $\dot{\epsilon}$ are related through the expression:

46

$$\sigma = B \dot{\epsilon}^{m} \qquad [3]$$

where m is the reciprocal of the stress exponent, n, in eqn. [1] and is termed the strain rate sensitivity, and B is a constant involving the temperature dependence. The values of m range from 0.004 for pure metals to 0.5 for superplastic materials and Class I solid solution alloys with m = 0.33 lie between the above two (13). An empirical correlation has been developed (14) between the total strains $\Delta L/L$ and the value of the strain rate sensitivity m ;

$$\frac{\Delta L}{L} = \exp\left(\frac{mK}{1-m}\right)^{-1} \qquad [4]$$

where K is typically 2-3. Using the data for WC-3009 and taking K = 2, the predicted strains from eqn.[4] agree quite well with the observed strains.

SUMMARY OF RESULTS

A commercial Nb-base alloy,WC-3009 has been successfully produced by P/M techniques. Results of the present investigation are as follows: (1) P/M alloy WC-3009 consolidated by hot pressing at 1590 K and 200 MPa pressure was fully dense. (2) Tensile strengths of the P/M alloy at 300-1590 K were similar to the arc-melted alloy. However P/M alloy showed less tensile elongation than the arc-melted alloy. (3) WC-3009 exhibited Class I solid solution behavior with n = 3.2 and Q = 270 kJ/mol. (4) WC-3009 exhibited high creep-rupture ductilities due to its high strain rate sensitivity (m = 0.3).

ACKNOWLEDGEMENTS

The authors would like to thank Dr. Hugh R. Gray, Chief, Advanced Metallics Branch of the NASA-Lewis Reseacrh Center for supporting this work through the HITEMP research grant.

REFERENCES

(1) M. G. Hebsur, J. R. Stephens, J. L. Smialek, C. A. Barrett, and D. S. Fox, " Influence of Alloying Elements on the Oxidation Behavior of NbAl3 ", NASA-TM 101398 NASA-LeRC Cleveland OH. , September 1988.

(2) R. H. Titran and R. W. Hall, " High Temperature Creep Behavior of a Columbium Alloy,FS85 " , NASA TN D2885, NASA-LeRC Cleveland OH. , June 1965.

(3) C. C. Wojcik, " Evaluation of Powder Metallurgy Processed Nb-30Hf-9W(WC-3009) " , in " Modern Developments in Powder Metallurgy ", Vol. 19, (Published by MPIF. APMI, Princeton NJ, June 1988), 187-200

(4) A. K. Mukherjee, J. E. Bird, and J. E. Dorn, " Experimental correlations for high temperature creep " , ASM Trans. Quarterly , 62(1969)155-79.

(5) F. Garofalo, " Fundamentals of Creep and Creep-Rupture in Metals " , (Macmillan, New York, NY. , 1965), 19-65

(6) F. E. Armstrong and H. L. Brown, " Anomalous Temperature Dependence of Elastic Moduli of Niobium " , Metal. Trans. ASM 58(1965)30-37.

(7) J. Wadsworth, C. A. Roberts and E. H. Rennhack, " Creep behavior of Hot Isostatically Pressed Niobium Alloy Powder Compacts " , J. Mat. Sci. ,17(1982)2539-46.

(7) J. Wadsworth, C. A. Roberts and E. H. Rennhack, " Creep behavior of Hot Isostatically Pressed Niobium Alloy Powder Compacts ", J. Mat. Sci. ,17(1982)2539-46.

(8) M. G. Hebsur and I. E. Locci, "Microstructure in Rapidly Solidified Niobium Aluminides ", NASA-TM 100264, NASA-LeRC, Cleveland OH. , March 1988.

(9) O. D. Sherby and P. M. Burke, " Mechanical Behavior of Crystalline Solids at Elevated Temperature ", Progress in Material Science , B. Chalmers and W. Hume-Rothery, eds. , Vol. 13, No. 7, (Pergamon Press, Oxford, 1967), 325-390.

(10) B. Ilschner and W. D. Nix, " Mechanisms controlling creep of single phase metals and alloys ", in " Strength of Metals and Alloys ", Proc. Of the 5th Int. Conf. (ICSMA 5), Aachen, West Germany, August 1979, P. Hassen et Al., eds. , Vol. 3, (Pergmon Press,Exeter,1980), 1503-1530.

(11) R. H. Titran and W. D. Klopp, " Long Time Creep Behavior of the Niobium Alloy C-103 ", NASA-TP 1727, NASA-LeRC Cleveland,OH. , Oct. , 1980.

(12) F. G. Arcella, " Interdiffusion Behavior of Tungsten or Rhenium and Group V and VI Elements and Alloys of the Periodic Table, Part I ", NASA-CR 134490, NASA-LeRC Cleveland, OH. , Sept. , 1974 .

(13) J. Wadsworth, T. Oyma and O. D. Sherby, " Super Plasticity: Prerequisite and Phenomenology ", in " Adv. Mater. Technol. Am. ,1980 ", 6th Inter Am. Conf. Mater. Technol. , San Francisco, August 1980, I. LeMay, ed. , (American Society of Mechanical Engineers, NewYork NY. , 1980), 29-41.

(14) F. A. Mohamed, " Modification of the Burke-Nix ductility expression ", Scripta Metallurgica, 13(1979)87-94.

TENSILE BEHAVIOR OF TUNGSTEN AND TUNGSTEN-ALLOY

WIRES FROM 1300 TO 1600 K

Hee Mann Yun*

National Aeronautics and Space Administration
Lewis Research Center
Cleveland, Ohio 44135

Summary

The tensile behavior of 200-μm-diameter tungsten lamp (218CS-W), tungsten + 1.0 atomic percent (a/o) thoria (ST300-W), and tungsten + 0.4 a/o hafnium carbide (WHfC) wires was determined over the temperature range 1300 to 1600 K at strain rates of 3.3×10^{-2} to 3.3×10^{-5} sec^{-1}. Although most tests were conducted on as-drawn materials, one series of tests was undertaken on ST300-W wires in four different conditions: as-drawn and vacuum annealed at 1535 K for 1 hr, with and without electropolishing. Whereas heat treatment had no effect on tensile properties, electropolishing significantly increased both the proportional limit and ductility, but not the ultimate tensile strength. Comparison of the behavior of the three alloys indicates that the HfC-dispersed material possesses superior tensile properties. Theoretical calculations indicate that the strength/ductility advantage of WHfC is due to the resistance to recrystallization imparted by the dispersoid.

*National Research Council – NASA Research Associate. NASA Lewis Research Center, Cleveland, Ohio 44135.

Refractory Metals: State-of-the-Art 1988
Edited by P. Kumar and R.L. Ammon
The Minerals, Metals & Materials Society, 1989

Introduction

In high temperature metal matrix composites, the reinforcing fiber carries the preponderance of the applied load; thus tungsten alloys have typically been chosen as the preferred fiber because of their great strength and stiffness at temperatures greater than 1300 K. In the past, high-temperature tensile and creep behavior (1,2) of as-drawn tungsten and tungsten-alloy wires have been studied. Their properties were attributed to potassium dopants (3), in the case of lamp grade tungsten (218CS-W) or thoria dispersoids (4) (ST300-W), as well as heavily drawn fibrous microstructures. In the present study the tensile properties of current commercial and experimental tungsten-alloy wires are examined over the temperature range 1300 to 1600 K (homologous temperature ranging from 0.4 to 0.5) as a function of strain rate. In addition, both as-drawn (cold worked) and annealed thoria-strengthened alloy wires were tested with and without an electropolished surface in order to determine the effect of processing on mechanical properties.

Experimental Procedure

Materials

Table I shows chemical compositions of the three tungsten-alloy wires examined in this work. The tungsten alloys 218CS-W (strengthened with potassium bubbles) and ST300-W (thoria-strengthened) were commercially available, whereas the HfC-strengthened material was an experimental alloy. In all cases the alloy wires were fabricated by powder metallurgy techniques and drawn to their final nominal diameter of 0.2 mm. The fibrous grain structure of these three materials is illustrated in Fig. 1. The ST300-W wires were examined both in the as-drawn condition and after a 1 hr vacuum stress relief anneal at 1535 K (Fig. 2(a)) which is typical of the processing heat treatments utilized to fabricate metal matrix composites (5). In addition, specimens electropolished in a 1 n NaOH solution at 9V for 30 sec with the wire being rotated at about 1 revolution/sec (6) were tested. Polishing reduced the specimen diameter to 0.14 mm with a mean deviation of 0.01 mm along the approximately 25.4-mm-long electropolished zone.

Table I. - Chemical Composition of 218CS-W,
ST300-W, and WHfC Wires

Material	Chemical composition, a/o				
	C	Hf	ThO_2	K	W
ST300-W	----	----	1.0	-----	Balance
WHfC	0.44	0.41	---	-----	Balance
218CS-W	----	----	---	0.038	Balance

Mechanical Property Test Procedures

Tensile testing was conducted in a vacuum of 10^{-4} Pa at temperatures ranging from 1300 to 1600 K. Tensile properties were determined at constant cross-head speeds ranging from 0.00085 to 0.85 mm/sec; furthermore, all tensile strength properties were determined from the autographically recorded load - time curves. Because the wire specimens were mechanically

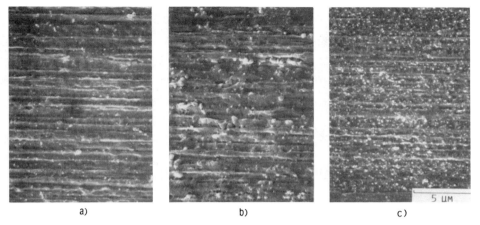

Figure 1. - Scanning electron microscope photomicrographs of as-drawn tungsten alloy wires, prior to tensile testing (longitudinally sectioned and etched), of a) 218CS-W, b) ST300-W, and c) WHfC.

Figure 2. - Scanning electron microscope photomicrographs of a) as-drawn and vacuum-annealed (for 1 hr at 1535 K) ST300-W wires, longitudinally sectioned and etched, and b) polished and unpolished surfaces of as-drawn ST300-W wires (all made prior to tensile testing).

gripped outside the hot zone of the furnace, the gauge length was not known; hence the data were analyzed under gauge-length-independent conditions (i.e., only the proportional limit, ultimate tensile strength, and reduction of area were determined). Tensile tests of electropolished samples were evaluated with the assumption that the deformation was taking place only in the reduced diameter gauge section zone. This supposition was reasonable, as shown by several tests measuring elongation of electropolished specimens by optical tracking with a cathetometer. These tests were conducted at the slowest velocity--an initial strain rate of 3.3×10^{-5} sec^{-1}. Good agreement was found between strain rates calculated from extension - time data, $(2.5 \pm 0.5) \times 10^{-5}$ sec^{-1}, and cross-head motion - time data, $(3.3 \pm 0.3) \times 10^{-5}$ sec^{-1}.

Microstructural Examination

The longitudinal grain structure of as-received, heat-treated, electropolished, and tested wires was studied with both light optical and scanning electron microscope (SEM) techniques.

Results

Influence of Surface Finish and Annealing on ST300-W

Typical tensile properties of the ST300-W wire, including proportional limit (PL), ultimate tensile strength (UTS), and reduction of area (RA) are shown in Fig. 3 as functions of temperature, surface condition, and heat treatment for materials tested at a constant cross-head speed of 0.0085 mm/sec (approximate strain rate of 3.3×10^{-4} sec^{-1}). Both strength (Fig. 3(a)) and ductility (Fig. 3(b)) generally decrease with increasing temperature. Although the elevated temperature UTS is essentially independent of surface finish and/or heat treatment, electropolishing has clearly improved the PL and RA. The 1535 K anneal for 1 hr appears to have no effect. The higher PL and RA of the electropolished specimens, compared to the as-drawn samples, are probably due to the elimination of surface flaws which cause (1) premature yielding by local reduction of the cross-sectional area and (2) premature fracture by joining the surface defects with internal cracks.

The microstructural difference between as-drawn and annealed ST300-W wire (Fig. 2(a)) was slight; namely (1) a partial destruction of the fibrous structure and (2) a slight increase in the width of the grains. The anneal can, therefore, be characterized as a stress relief heat treatment wherein the grain and subgrain structure remain essentially unchanged. This behavior is in agreement with Tajime's (7) view that complete recovery in tungsten wires can only begin to take place above 1573 K. Therefore, it can be concluded that the heat treatment at 1535 K does not cause any strength or ductility decrease in the temperature range from 1300 to 1600 K.

Tensile Properties

The influence of alloy chemistry on the PL and UTS is shown in Fig. 4 where these properties are plotted as a function of temperature for materials tested at a constant cross-head speed of 0.0085 mm/sec (approximate strain rate of 3.3×10^{-4} sec^{-1}). From 1300 to 1600 K, the hafnium carbide-dispersed WHfC wires consistently displayed higher PL's (Fig. 4(a)) than either the thoria- or potassium bubble-dispersed wires. Below 1500 K the thoria-dispersed ST300-W possessed higher yield strengths

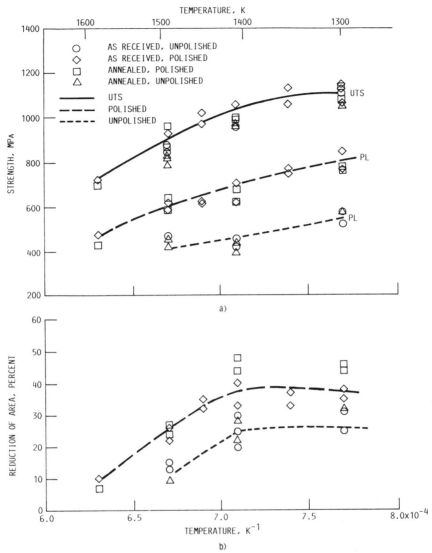

Figure 3. - Effect of surface condition and a 1-hr vacuum anneal at 1535 K on a) ultimate tensile strength and proportional limit and b) reduction of area for ST300-W wires tested at 0.0085 mm/sec (estimated initial strain rate = 3.3×10^{-4} sec^{-1}).

than potassium bubble-dispersed wires; however, with increasing temperature the difference in yield strengths becomes small. In terms of ultimate tensile strength, the WHfC wire has a considerable strength advantage over the 218CS-W or ST300-W at the higher test temperatures (Fig. 4(b)). However, at 1300 K the strength of WHfC and ST300-W are similar, and the potassium bubble-strengthened wire 218CS-W is much weaker.

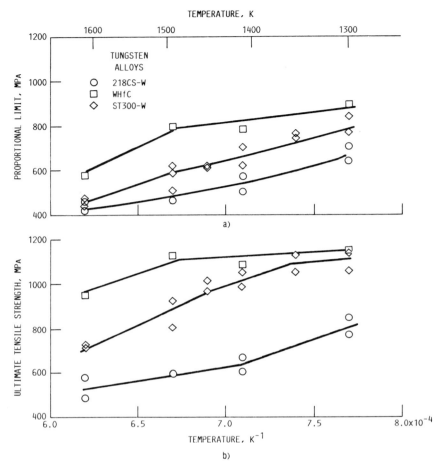

Figure 4. - a) proportional limit and b) ultimate tensile strength of tungsten-alloy wires tested at 0.0085 mm/sec.

Stress - Strain Rate Behavior

The effect of strain rate on the PL and RA is shown in Figs. 5 and 6, respectively, for as-drawn, electropolished wires at 1400 and 1600 K. The measure of yield strength is relatively constant over two to three orders of magnitude of strain rate for WHfC wires at 1400 K (Fig. 5(a)) and for ST300-W and 218CS-W at 1600 K for fast strain rates (Fig. 5(b)). Conversely, the PL's gradually decrease for WHfC wires at 1600 K (Fig. 5(b)) and ST300-W and 218CS-W wires at 1400 K (Fig. 5(a)) as the strain rate is decreased from 0.033 to 0.000033 sec^{-1}. By following normal convention, the data in Fig. 5 have been analyzed in terms of the power law deformation behavior,

$$\dot{\epsilon} = A\sigma^n \qquad (1)$$

$$\sigma = B\dot{\epsilon}^m \qquad (2)$$

54

where $\dot{\varepsilon}$ and σ are the strain rate and stress (PL in this work), respectively, n is the stress exponent, and m is the strain rate sensitivity with n = 1/m. Table II summarizes the calculated strain rate sensitivities as functions of temperature, strain rate range, and alloy composition. Warren et al. (2) reported n = 7 to 8 (m = 0.13 to 0.14) for ST300-W-like thoria-dispersed tungsten wires tested at 1400 K and a strain rate of 10^{-8} sec^{-1}. This is somewhat different from m = 0.086 (n = 11.5), determined for ST300-W at 1400 K and strain rates less than 10^{-3} sec^{-1}. The different stress exponent would be caused from the different strain rate region.

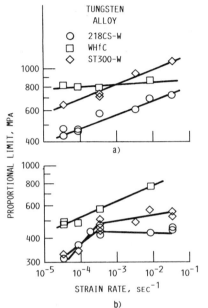

Figure 5. - Proportional limit at a) 1400 K and b) 1600 K as a function of initial strain rate for several tungsten-alloy wires.

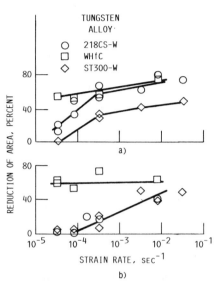

Figure 6. - Reduction of area at a) 1400 K and b) 1600 K as a function of initial strain rate for several tungsten-alloy wires.

Table II. - Strain Rate Sensitivities of Various W-Alloy Wires

Material (dispersoid)	Temperature, K	Strain rate sensitivity, m	Strain rate range, sec^{-1}
ST300-W (thoria)	1400	0.086	3×10^{-5} to 3×10^{-2}
	1600	.19	3×10^{-5} to 1×10^{-3}
	1600	.039	1×10^{-3} to 3×10^{-2}
218CS (bubbles)	1400	.086	3×10^{-5} to 3×10^{-2}
	1600	.15	3×10^{-5} to 1×10^{-3}
	1600	.001	1×10^{-3} to 3×10^{-2}
WHfC (HfC)	1400	.014	3×10^{-5} to 3×10^{-2}
	1600	.091	3×10^{-5} to 3×10^{-2}

In general, hafnium carbide-dispersed WHfC wires show a higher fracture
ductility both at 1400 K (Fig. 6(a)) and 1600 K (Fig. 6(b)) than either the
bubble-dispersed 218CS-W or the thoria-dispersed ST300-W. The ductility of
WHfC is nearly constant --about 70 percent-- whereas that of 218CS and
ST300-W at both 1400 and 1600 K drops from about 60 percent to 5 percent as
the strain rate decreases.

Microstructure of Tested Wires

The microstructure of longitudinal sections and the fracture morphology
of ST300-W wire after testing at 1600 K are shown in Fig. 7. Clearly, the
fracture mode is dependent on deformation rate since essentially no necking
was observed after slow straining (Fig. 7(a)); significant necking in the
failure area was found after fast strain rate testing (Fig. 7(c)). Instead
of failing by plastic flow, the slowly deformed samples apparently failed by
the formation and growth of surface cracks along grain boundaries, with the
general direction of the cracks being perpendicular to the tensile stress
axis. Although the initial substructure of ST300-W (Fig. 1(b)) has been
maintained in the high strain rate specimen (Fig. 7(d)), it has been
segmented into 4- to 15-μm lengths, and its width has significantly
increased after slow testing (Fig. 7(b)). The fracture behavior of ST300-W
at 1400 K was identical to that observed at 1600 K. The WHfC fractures
after fast and slow straining at either 1400 or 1600 K were analogous to
those found after fast testing of ST300-W (Fig. 7(c)). The post-test
microstructure of 218CS-W exhibited effects of strain rates on the fracture
characteristics similar to those of ST300-W.

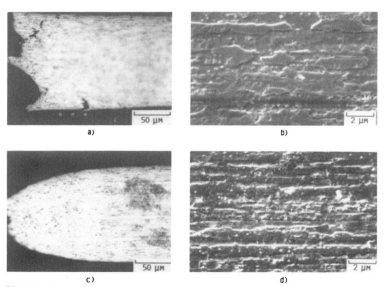

a)

b)

c)

d)

Figure 7. - Scanning electron microscope and light photo-
    micrographs of ST300-W wires tested at 1600 K and a strain
    rate of 0.000033 sec^{-1} a) fracture region and b) away
    from fracture; and at a strain rate of 0.033 sec^{-1} c)
    fracture region and d) away from fracture.

By utilizing an intersection method (lines perpendicular to the wire axis), the width of the fibrous grain structure **t** was determined. For each condition, nearly 200 measurements were made at a distance about 1 mm from the fracture surface. Figure 8 illustrates the relative frequency of **t** as a function of strain rate for ST300-W and WHfC tested at 1600 K. Clearly, the

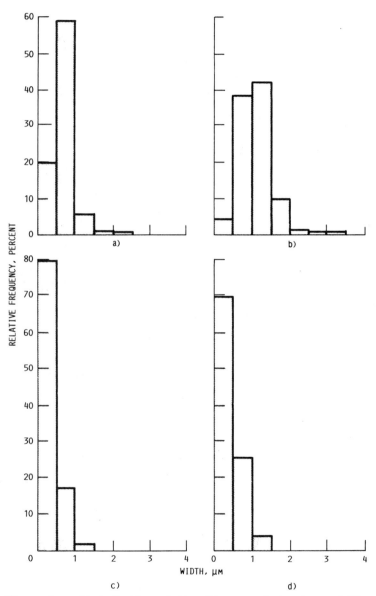

Figure 8. - Distribution of the fibrous substructure width, t, as a function of strain rate tested at 1600 K: a) 3.3×10^{-2} sec^{-1} for ST300-W, b) 3.3×10^{-5} sec^{-1} for ST-300-W, c) 3.3×10^{-2} sec^{-1} for WHfC, and d) 3.3×10^{-5} sec^{-1} for WHfC.

average grain width of ST300-W increased with decreasing strain rate (the median value of **t** was shifted from the 0.5 to 1.0 μm range to the 1.0 to 1.5 μm range). However, the median value of **t** for WHfC was unchanged under the same conditions. Since these microstructural observations can be correlated with the relationship between ductility and strain rate (Fig. 6(b)), losses in ductility might stem from microstructural instabilities owing to the original substructure being replaced due to drawing during the tensile tests.

Discussion

Strengthening Mechanisms

The tensile strength of a material can be increased by introducing a dispersion of second particles and/or by cold working. For example, particles which impede the motion of dislocations increase the Orowan stress. For a given matrix, the type of particle and the average diameter have some influence on the Orowan stress; however, the major factor affecting its magnitude is particle spacing. Dislocation substructure formed by prior cold working, where the stored elastic energy, density of sessile dislocations, and the average diameter of the subgrains affect strength, also impedes the movement of mobile dislocations. By utilizing data from the literature, the expected Orowan stresses for different dispersoids in the tungsten-alloy wires can be calculated and compared to the anticipated strengthening due to the retained fine substructure.

Strength Contribution via the Orowan Stress. The theoretical shear yield stress of dispersion-strengthened alloys has been calculated by Ashby (8). For a given second-phase diameter **d** and interplaner particle spacing distance **D**, the Orowan yield stress Γ_O on the edge dislocations is directly proportional to the matrix shear modulus G_m (9); that is

$$\Gamma_O = (G_m b/2.36\pi D)\ln(d/2b) \tag{3}$$

where **b** is the magnitude of the Burgers vector. With the assumption that the matrix chemistry is similar for all three wires, G_m and **b** are assumed to be identical for each system. Therefore, from previously measured values of **d** and **D** (4,10,11) (Table III), the Orowan stress for each type of wire can be calculated. For instance, at 1400 K Γ_O equals 68 MPa for 218CS-W, 80 MPa for ST300-W, and 118 MPa for WHfC wires. The WHfC alloy is expected to be somewhat stronger than the others due to a higher Orowan stress.

Table III. – Characterization of the Dispersoids in the W-Alloy Wires

Dispersoid	Fiber	Average second-phase diameter, d, nm	Average interplaner particle distance, D, nm	Second-phase volume fraction, f, percent	Particle shear modulus, G_p,[a] G_m	Reference
Thoria	ST300-W	76	315	3.80	G_m	4
HfC	WHfC	35	180	1.55	G_m	10
Bubble	218CS-W	15	250	0.19	0.1 G_m	11

[a]In terms of G_m, the tungsten matrix shear modulus.

Substructure Strengthening. One contribution of the substructure to the yield stress results from the stored elastic energy, which can be considered to be a direct function of prior work-hardening. Such work-hardening is reported to be dependent upon the volume fraction of the second-phase **f** (8,12) as well as the accumulated plastic shear strain **a** (12). Work-hardening, derived from the dispersed second phase, is a result

58

of interactions between moving dislocations and particles according to Ashby (8). Thus the strengthening due to work-hardening Γ_{WH} is proportional to the root of the accumulated plastic strain,

$$\Gamma_{WH} = 0.24 G_m \sqrt{bfa/d} \qquad (4)$$

Unfortunately the value of **a** cannot be determined presently because it is a direct function of the complicated and undefined fabrication schedules for each type of wire. If it is assumed that during production the original 10-μm grain diameter is reduced to 0.2 μm, the plastic shear strain would be 15.6 --2ℓn(10/0.2)^2-- and the shear stresses required for further deformation would be

$$\Gamma_{WH} = 343 \text{ MPa for 218CS-W}$$
$$\Gamma_{WH} = 1530 \text{ MPa for ST300-W}$$
$$\Gamma_{WH} = 1386 \text{ MPa for WHfC}$$

These values are most likely overestimates as some recovery occurs during process anneals which are part of wire fabrication schedules. The thoria- and HfC-strengthened materials should be much stronger than the potassium bubble-dispersed alloy. The work-hardening in a second-phase-strengthened alloy is estimated by Brown et al. (13) by a mean internal stress σ_M, which is proportional to the accumulated plastic strain in the matrix;

$$\sigma_M = 2Ka(G_p G_m f)/[G_p - K(G_p - G_m)] \qquad (5)$$

where K is a constant, ranging from 0.5 to 0.78 (14) depending on the accommodation between the matrix and particle. The required shear stress can be calculated if we assume that G_p is the particle shear modulus, that $G_p = G_m$ for thoria and hafnium carbide, and that $G_p = 0.1 G_m$ for bubbles.

Calculated Orowan stresses and substructure strengths, normalized with respect to shear modulus of the matrix, are summarized in Table IV. For computations utilizing either Ashby's or Brown's model, three different values of the accumulated plastic shear strain were assumed: **a** = 0.1, 0.5 and 15.6. With the exception of the bubble-strengthened alloy, the Orowan stresses are generally small in comparison to the strength due to prior work. Clearly, Brown's substructure model indicates large differences in strength between bubble- and particle-hardened tungsten alloys, whereas Ashby's model is much less sensitive to the type of second-phase particle. For either model the strengths of the HfC or thoria alloys are similar; Brown's expression, however, leads to greater strengths than Ashby's.

Table IV. - Expected Orowan Stress Due to Dispersoids and Substructure
Stress Due to Prior Working

Material	Orowan-to-matrix stress,[a] Γ_0/Γ_m	Work-hardening-to-matrix stress,[b] Γ_{WH}/Γ_m, 10^{-4}						Orowan-to-work-hardening stress,[b] Γ_0/Γ_{WH}	
		a = 0.1		a = 0.5		a = 15.6		a = 0.5	
		Eq. 4	Eq. 5	Eq. 4	Eq. 5	Eq. 4	Eq. 5	Eq. 4	Eq. 5
218CS-W	4.89×10^{-4}	4.47	1.27	10	6.3	55.8	198	0.33	0.44
ST300-W	5.78	8.91	38	19.9	190	111	5930	0.23	0.03
WHfC	8.53	8.36	15.5	18.7	77.5	104	2420	0.31	0.10

[a]From Eq. (3).
[b]For various a, accumulated plastic shear strain, mm/mm.

Figure 9 compares the measured to the estimated theoretical shear yield stress as functions of temperature and accumulated plastic shear strain. In this plot both shear strengths have been normalized with respect to shear modulus for tungsten where the experimental data were taken at a shear strain rate of 6.6×10^{-4} sec^{-1}. The theoretical values were based on a superposition of each strengthening mechanism where

$$\Gamma_{tot} = \Gamma_0 + \Gamma_{WH} + \Gamma_m \qquad (6)$$

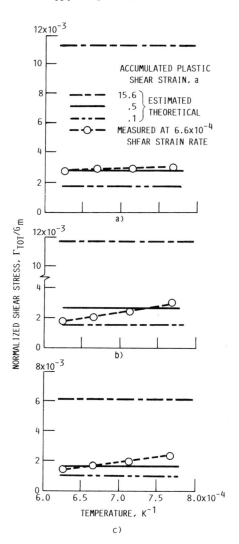

Figure 9. – Effect of temperature and prior work on theoretical shear stress due to superposition of substructure free motion, Orowan stress, and work-hardening stress for a) WHfC, b) ST300-W, and c) 218CS-W.

The matrix shear stress Γ_m, which is considered to be identical for all three wires, was taken as the strength of tungsten in the substructure-free and equiaxed grain size condition. The theoretical substructure stress was estimated from the Ashby model (Eq. (4)) because this model is the least sensitive to the plastic shear strain value. The estimated shear stresses for the wires as a function of temperature are in good agreement with the experimental data (shear stress = PL/2), if **a** is assumed to be about 0.5 for both ST300-W and WHfC. For 218CS-W a higher **a** value might be required.

Stability of the Fibrous Structure

The sharply decreasing strength of ST300-W and 218CS-W at strain rates less than 10^{-4} sec^{-1} at 1400 K (Fig. 5(a)) and at 1600 K (Fig. 5(b)) could be due to lack of second-phase stability and concurrent loss of resistance to recrystallization. According to Snow (15), Davis (16), and Barna (17), primary static recrystallization to a polycrystalline structure occurs in the temperature range of 1300 to 1400 K in 0.2 to 10 hr and is accompanied by a loss of the preferred fiber texture and a decreased grain aspect ratio. Secondary recrystallization, on the other hand, typically takes place above 1600 K with the loss of the fibrous structure and the coarsening of the doped or alloyed tungsten grains. During hot tensile or creep test conditions these temperature/time parameters should be reduced since dynamic recrystallization is an easier process.

The inhibition of recrystallization by a second phase is attributed to particles impeding the motion of the recrystallization front (11,18). Recrystallization will take place if the driving force of the grain boundary mobility is larger than the inhibiting force. According to Warlimont et al. (18) the impeding stress P_i is dependent on the second-phase distribution;

$$P_i = (2\tau_B f_B / r) + (2\tau_{SB} f_{SB} / r) \qquad (7)$$

where τ_B and τ_{SB} represent specific grain or subgrain boundary energy, f_B and f_{SB} are the volume fraction of second-phase particles intersected by a unit of grain or subgrain, and r is the radius of particles or bubbles.

The driving force for recrystallization P_d is a result of the existing grain and subgrain boundary structure;

$$P_d = (2\tau_B / X_B) + (2\tau_{SB} / X_{SB}) \qquad (8)$$

where X_B and X_{SB} are grain diameter and subgrain spacing, respectively. By equating Eqs. (7) and (8), the critical radius to recrystallization is defined

$$r_c = (1.1 X_B X_{SB} f_B) / (X_{SB} + 0.1 X_B) \qquad (9)$$

when it is assumed that $\tau_{SB} = 0.1\tau_B$ and $f_{SB} = f_B$ (11). The theoretical equilibrium state for the radius of second-phase particles can be obtained by using parameters from the literature (18) where $\tau_B = 1.08$ J/m^2, $\tau_{SB} = 0.1$ J/m^2, $X_B = 0.5$ µm, $X_{SB} = 0.15$ µm, and $f_B = 5\pi d^2/4D^2$ for a grain boundary mobility perpendicular to the fiber axis; thus

$$r_c, \text{bubble} = 6 \text{ nm}$$
$$r_c, \text{thoria} = 90 \text{ nm}$$
$$r_c, \text{HfC} = 60 \text{ nm}$$

Since these critical radii are the thresholds to recrystallization, larger values prohibit recrystallization.

Comparing the calculated radii to measured parameters (Table III) reveals that the estimated critical radii of thoria and HfC are greater than the initial particle radii, whereas the estimated critical size for the potassium-filled bubbles is less than that for the as-received material. Hence, on the basis of this theoretical calculation, one would expect bubble-strengthened materials to undergo recrystallization rather easily, whereas both particle-strengthened alloys should be resistant to such phenomena. This would seemingly indicate that the HfC- and thoria-containing materials should possess mechanical properties with less temperature and time sensitivity. In general, this seems to be the case (see Figs. 3 and 4).

If particle growth takes place during elevated temperature exposure, it is probable that the grain boundaries could break free and recrystallization would take place. Thus it becomes a question of which type of second phase is more stable. Unfortunately, the basic information, such as solubility of the elements comprising the particles in the metal matrix, interfacial energy between particle and matrix, and diffusivity, is not known; hence computations based on simple models (i.e., Wagner (19)) cannot realistically be made. However, based on the current work (Figs. 4 and 5), HfC appears to be more resistant to recrystallization than thoria.

Summary of Results

The tensile properties of 200-μm lengths of 218CS-W, ST300-W, and WHfC wires were examined in the temperature range of 1300 to 1600 K, and the following results were obtained:

1. A stress relief at 1535 K for 1 hr did not affect the tensile strength or ductility of ST300-W.

2. Electropolishing of ST300-W significantly improved the proportional strength and ductility but not the ultimate tensile strength.

3. The hafnium carbide-dispersed alloy had significantly better tensile properties than either the thoria- (ST300) or potassium bubble- (218CS) strengthened materials.

4. Calculations indicate that hafnium carbide-dispersed (WHfC) wires should possess a higher Orowan stress and a stable fibrous substructure and, thus, have superior high temperature tensile properties.

5. Hafnium carbide-dispersed (WHfC) wires are less sensitive to strain rate than bubble- or thoria-dispersed wires.

Conclusions

From the current work it is concluded that using hafnium carbide-strengthened wires will lead to stronger metal matrix composites than those from either of the currently available commercial tungsten alloys. The properties of hafnium carbide wires can be further improved by electropolishing to remove surface defects prior to composite fabrication.

Acknowledgments

The author wishes to thank Dr. J.D. Whittenberger for extensive discussions and Dr. R.H. Titran and D.W. Petrasek for support and encouragement.

References

1. B. Harris and E.G. Ellison: Trans. ASM, 1966, vol. 59, pp. 744-754.

2. R. Warren and C-H. Anderson: Proc. 10th Plansee Seminar, Vol. 2, H.M. Ortner, ed., Metallwerk Plansee, Austria, 1981, pp. 243-246.

3. D.B. Snow: Met. Trans. A, 1979, vol. 10, pp. 815-821.

4. G.W. King: Trans. TMS-AIME, 1969, vol. 245, pp. 83-89.

5. L.J. Westfall, D.W. Petrasek, D.L. McDanels, and T.L. Grobstein: NASA TM-87248, National Aeronautics and Space Administration, Washington, DC, 1986.

6. L.H. Amra, L.F. Chamberlain, F.R. Adams, J.G. Tavernelli, and G.J. Polanka: NASA CR-72654, National Aeronautics and Space Administration, Washington, DC, 1970.

7. Behavior and Properties of Refractory Metals, T.E. Tietz and J.W. Wilson, eds., Stanford University Press, Stanford, CA, 1965, p. 295.

8. M.F. Ashby: in Oxide Dispersion Strengthening, G.S. Ansell, T.D. Cooper, and F.V. Lenel, eds., Science Publishers, New York, 1968, pp. 143-205.

9. P.E. Armstrong, and H.L. Brown: Trans. TMS-AIME, 1964, vol. 230, pp. 962-966.

10. G.W. King and D.W. Petrasek: NASA TM-79115, National Aeronautics and Space Administration, Washington, DC, 1979.

11. H.P. Stuewe: Met. Trans. A, 1986, vol. 17, pp. 1455-1459.

12. R. Ebeling, and M.F. Ashby: Philos. Mag., 1966, vol. 13, pp. 805-834.

13. L.M. Brown, and D.R. Clarke: Acta Metall., 1977, vol. 25, pp. 563-570.

14 L.M. Brown, and D.R. Clarke: Acta Metall., 1975, vol. 23, pp. 821-830.

15. D.B. Snow: Met. Trans. A, 1976, vol. 7, pp. 783-794.

16. G.L. Davis: Metallurgia, 1958, vol. 58, pp. 177-184.

17. A. Barna, I. Gaal, O. Geszti-Herkner, G. Radnoczi, and L. Uray: High Temp.-High Press., 1978, vol. 10, pp. 197-205.

18. H. Warlimont, G. Necker, and H. Schultz: Z. Metall., 1975, Vol. 66, pp. 279-286.

19. A.U. Seybolt: in Oxide Dispersion Strengthening, G.S. Ansell, T.D. Cooper, and F.V. Lenel, eds., Science Publishers, New York, 1968, pp. 469-487.

HIGH-TEMPERATURE MECHANICAL PROPERTIES OF W-Re-HfC ALLOYS

B. L. Chen, A. Luo, K. S. Shin, and D. L. Jacobson

Department of Chemical, Bio and Materials Engineering
Arizona State University
Tempe, AZ 85287, U. S. A.

Abstract

Tungsten-base alloys have great potential for future high-power space systems because of their ultimate thermal capabilities and exceptional high-temperature mechanical properties. However, the intrinsic brittleness of tungsten-base alloys at low temperatures has limited their development and widespread application. The addition of rhenium and hafnium carbide is known to improve low-temperature ductility and high-temperature strength of tungsten alloys. In the present study, high-temperature tensile and creep properties of tungsten-rhenium-hafnium carbide (W-Re-HfC) alloys were investigated over a wide temperature range to examine the strengthening effect of hafnium carbide and the high-temperature deformation mechanisms. The change in microstructure was also examined. It is found that the finely dispersed hafnium carbide particles significantly improve the high-temperature tensile and creep properties of W-Re-HfC alloys. The stress exponent for high-temperature creep is found to be 5.2 and the activation energy for creep deformation is 105 Kcal/mole in the temperature range of 1955 to 2190 K. Dislocation core diffusion is considered to be the dominant factor for the creep deformation of W-Re-HfC alloys.

Refractory Metals: State-of-the-Art 1988
Edited by P. Kumar and R.L. Ammon
The Minerals, Metals & Materials Society, 1989

Introduction

Tungsten has the highest melting point among all metals and thus possesses a number of desirable properties for high-temperature applications, such as excellent high-temperature modulus and strength, high emissivity and low vapor pressure. However, difficulties in processing and fabrication of tungsten due to its high melting point and intrinsic brittleness at low temperatures have limited its widespread application. In an attempt to overcome some of the difficulties, there was a significant effort at NASA Lewis Research Center during the Apollo era to develop tungsten-base alloys with improved low-temperature fabricability and high-temperature mechanical properties for space applications (1-9). It has been found that a moderate addition of rhenium can improve low-temperature ductility and high-temperature strength of tungsten (2,4,7). It has also been found that the presence of finely distributed hafnium carbide particles can dramatically increase the high-temperature strength of tungsten and tungsten-rhenium alloys (3,5,6).

In recent years, there has been a renewal of interest in the development of tungsten-base alloys for future space power system applications (10). There has also been an attempt to develop metal-matrix composites reinforced with tungsten fibers for structural applications in advanced aerospace propulsion systems. Tungsten-rhenium-hafnium carbide (W-Re-HfC) alloys are obvious candidate materials for such applications since they have both excellent high-temperature mechanical properties and good low-temperature ductility.

The previous study of Klopp and Witzke (5) has concluded that the optimum HfC content for maximum strengthening at elevated temperatures is 0.35 mol.% for W-4% Re alloys. Therefore the main objective of the present study is to examine both tensile and creep properties of W-Re-HfC alloys with similar compositions to W-4% Re-0.35 mol.% HfC over a wide temperature range above 0.5 T_m (melting point in Kelvin) in order to determine the strengthening effect of HfC and the high-temperature deformation mechanisms.

Experimental Procedure

Materials

The starting materials for the present study were two 25.4 mm diameter tungsten-rhenium-hafnium carbide (W-Re-HfC) rods provided by NASA Lewis Research Center. Both rods were processed by an arc-melting technique. One of the rods was received as arc-melted and the other was slightly swaged. The chemical compositions of both rods were determined by microprobe analysis and found to be tungsten-3.6 at.% rhenium-0.4 mol.% hafnium carbide (W-3.6Re-0.4HfC) for the as arc-melted rod and tungsten-4.0 at.% rhenium-0.33 mol.% hafnium carbide (W-4.0Re-0.33HfC) for the swaged rod. The concentration of carbon was slightly in excess of that of hafnium in both alloys so that all hafnium atoms could be assumed to exist in the combined form of hafnium carbide (HfC). Plate-type specimens for tensile and creep tests were machined with an electric discharge machine from W-3.6Re-0.4HfC and W-4.0Re-0.33HfC, respectively. The gauge lengths of the specimens for tensile and creep tests were 8.0 mm and 12.7 mm, respectively. All specimens were first mechanically polished with 400 and 600 grit SiC papers and then chemically polished with a 10% NaOH solution.

Mechanical Tests

Both tensile and creep tests were conducted in the specially designed ultra-high vacuum (UHV) test chambers mounted on the testing machines. The test chambers were evacuated with a mechanical pump, followed by a cryogenic sorption pump, and finally with an ion pump. Both tensile and creep test chambers have provisions to heat the specimens with self-resistant heating by passing an electric current through the specimen. The temperature of the specimen was measured with an optical pyrometer calibrated with a ribbon filament lamp prior to testing. The maximum uncertainty of the measured temperature was within 5 K over the entire temperature range of the present study.

All tensile tests were performed on an Instron testing machine in a vacuum better than 1.3×10^{-5} Pa. In order to provide a uniform recrystallized microstructure, all tensile specimens were initially annealed at 2350 K for 30 min. prior to testing. The strain rate employed was 10^{-3}/sec for all tensile tests. Both step-load and step-temperature creep tests were performed on a custom-made creep testing machine in a vacuum better than 10^{-4} Pa. All creep specimens were annealed initially at 1273 K for 12 hours for degassing and subsequently at 2438 K for 1.5 hours for recrystallization. The specimens were then kept at the desired test temperature for 30 min. prior to testing. The load was applied to the specimen through a bellows and the elongation of the specimen was measured with a linear variable differential transducer. Step-load tests were conducted in the temperature range of 1955 to 2783 K and in the stress range of 10 to 70 MPa. Step-temperature tests were performed at a constant stress of 42.3 MPa in the temperature range of 2013 to 2190 K. The side and fracture surfaces of the tested specimens were examined with an optical microscope and a JEOL 840 scanning electron microscope.

Experimental Results

Tensile Tests

The effect of test temperature on the 0.2% offset yield strength of W-3.6Re-0.4HfC above 0.5 T_m (1842K) is shown in Fig. 1. The results of

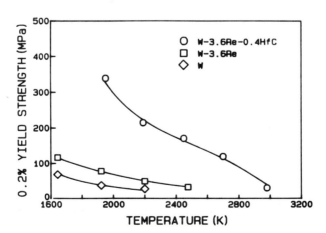

Figure 1 - Effect of temperature on the yield strength of W-3.6Re-0.4HfC.

arc-melted pure W and W-3.6Re reported by Klopp et al. (1,2) are also shown in the same figure for comparison. It can be clearly seen that the yield strength of W-3.6Re-0.4HfC is much greater than that of pure W or W-3.6Re up to 2700 K.

Fig. 2 shows the effect of test temperature on the ultimate tensile strength of W-3.6Re-0.4HfC. The comparison of the result with that of pure W or W-3.6Re again shows the dramatic effect of 0.4 mol.% HfC on the strength of W alloys up to 2700 K. The tensile strength of W-3.6Re-0.4HfC decreases rapidly above 2700 K and becomes similar to that of pure W or W-3.6Re at 2980 K.

In Fig. 3, the tensile elongation of W-3.6Re-0.4HfC is plotted as a function of temperature. The tensile elongation increases slowly with increasing temperature to 2450 K and then rapidly increases above 2450 K. The fracture surfaces of the tested specimens were examined with a

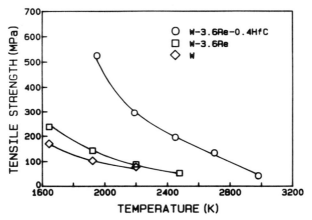

Figure 2 - Effect of temperature on the ultimate tensile strength of W-3.6Re-0.4HfC.

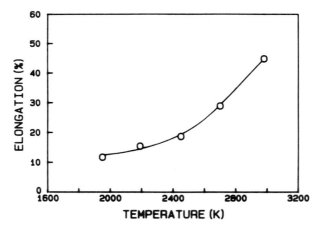

Figure 3 - Effect of temperature on the tensile elongation of W-3.6Re-0.4HfC.

scanning electron microscope. All specimens exhibited a typical ductile dimple fracture in the temperature range employed in this study. Figures 4 (a) and (b) show the scanning electron micrographs of the fracture surfaces of the specimens deformed at 1950 K and 2450 K, respectively. Although both specimens show a similar fracture mode, it can be seen from these fractographs that the tendency for ductile tearing increases with increasing temperature.

Creep Tests

Fig. 5 shows the typical strain-time creep curves of W-4.0Re-0.33HfC obtained from the step-load creep tests at 1955, 2190 and 2783 K with a successive increase in the applied stress. Upon initial loading, these

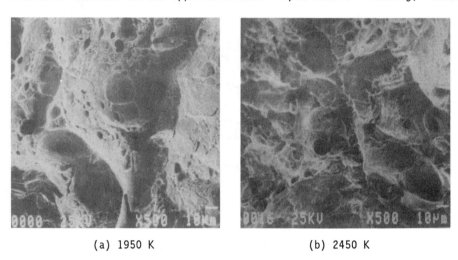

(a) 1950 K   (b) 2450 K

Figure 4 - The SEM fractographs of W-3.6Re-0.4HfC deformed at (a) 1950 K and (b) 2450 K.

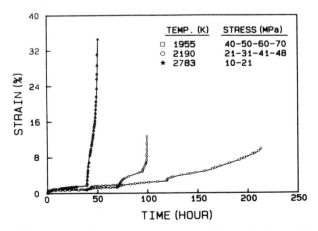

Figure 5 - Strain-time creep curves of W-4.0Re-0.33HfC obtained from the step-load tests at various temperatures.

69

curves all show a primary creep region with decreasing creep rate followed by a secondary creep region with a constant creep rate. In the secondary creep region, there is a balance between the strain-hardening rate and the recovery rate. An increase in the applied stress shifts this balance and the primary creep takes place again. Eventually tertiary creep occurs with necking in the gauge section followed by creep rupture (the creep test at 1955 K was terminated before rupture took place).

Fig. 6 shows a power law relationship between the steady-state creep rate of W-4.0Re-0.33HfC and the applied stress at various temperatures. The stress exponent n in the power law creep equation can be determined from the inverse of the slope of each straight line and is found to be in the range of 5.0 to 5.4 in the temperature range of 1955 to 2783 K.

The creep strength of W-4.0Re-0.33HfC at a creep rate of 10^{-6}/sec is compared with those of other W-base alloys (8,11-13) in Fig. 7. This

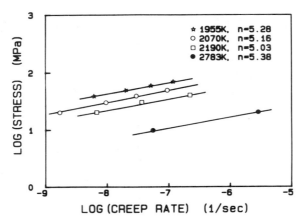

Figure 6 - Power law relationship between the steady-state creep rate of W-4.0Re-0.33HfC and the applied stress.

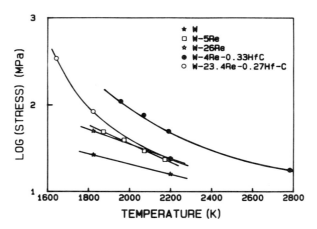

Figure 7 - Comparison of the creep strength at a creep rate of 10^{-6}/sec for various W-base alloys.

figure clearly shows the excellent creep strength of W-4.0Re-0.33HfC. For example, the comparison of the data shows that W-4.0Re-0.33HfC is about three times stronger than W-5Re at 1973 K. At 2200 K W-4Re-0.33HfC is about two times stronger than W-5Re, W-26Re, or W-23.4Re-0.27Hf-C.

The fracture of all creep specimens was initiated by intergranular cracking. Fig. 8 shows a typical metallographic picture of the side surface of the creep specimen after test. Although this particular picture was taken from the creep specimen tested at 2190 K, the creep specimen tested at other temperatures showed similar features. The picture shows typical wedge-shape intergranular cracks initiated by grain boundary sliding. It can be concluded, therefore, the creep rupture is initiated by grain boundary sliding.

Discussions

The excellent high-temperature strength of W-3.6Re-0.4HfC, as shown in Figs. 1 and 2, came from both the solid-solution strengthening of rhenium and the precipitation strengthening of hafnium carbide. However, the results of Klopp et al. (1,2) clearly indicated that the addition of 3.6 at.% Re to pure W only slightly improved the yield or ultimate tensile strength above 0.5 T_m. Therefore, it can be concluded that the superior strength of W-3.6Re-0.4HfC above 0.5 T_m is due to the addition of a small amount of HfC (0.4 mol.%). The finely dispersed HfC particles probably act as effective barriers to dislocation motion and improve the high-temperature strength up to 2700 K. The rapid decrease in high-temperature strength above 2700 K is most likely caused by the coarsening of HfC particles. Klopp and Witzke (5) also reported that coarsening of HfC particles took place above 2273 K.

The comparison of Figs. 1 and 2 shows that the difference between the yield strength and the ulimate tensile strength of W-3.6Re-0.4HfC decreases with increasing temperature. The difference becomes very small above 2700 K. This is most likely caused by a decrease in work hardening rate with increasing temperature. The work hardening behavior of a material can be expressed by the strain-hardening exponent n in the equation $\sigma = K\epsilon^n$, where σ and ϵ indicate true stress and true strain,

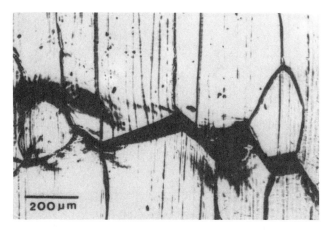

Figure 8 - Wedge-shape intergranular cracks on the side surface of the creep specimen tested at 2190 K.

repectively. The constant K is known as the strength coefficient of a material. The strain-hardening exponent n is a function of material, microstructure, temperature and strain rate. A large value of n indicates high work hardening capability and therefore high resistance of a material to further plastic deformation. Fig. 9 shows the effect of temperature on the strain-hardening exponent of W-3.6Re-0.4HfC. Unfortunately, the strain-hardening exponent of arc-melted pure W or W-3.6Re is not available in the literature. In Fig. 9, the result obtained from sintered and recrystallized W-3Re (14) is plotted for comparison. The strain-hardening exponent of W-3.6Re-0.4HfC rapidly decreases from 0.48 to 0.17 as the temperature increases from 1950 to 2450 K and then decreases slowly to 0.15 at 2980 K. The comparison of data indicates that the strain-hardening exponent of W-3.6Re-0.4HfC is much higher than that of W-3Re up to 2450 K and both alloys show a similar value of n above 2473 K. The formation of Orowan dislocation loops around finely distributed HfC particles is considered to be responsible for the high strain-hardening exponent of W-3.6Re-0.4HfC up to 2450 K.

As can be seen in Figs. 4 (a) and (b), the fracture mode of W-3.6Re-0.4HfC alloy above 0.5 T_m is ductile dimple tearing. The tendency for ductile tearing increases with increasing temperature. This tendency was also confirmed from the fractographs of the specimens deformed at other temperatures. The examination of the side surfaces of the deformed specimens revealed that extensive grain boundary sliding took place in the specimens tested at 2700 and 2980 K. Therefore, the rapid increase in tensile elongation above 2450 K is considered to be caused by the grain boundary sliding. It is noteworthy that the rapid increase in tensile elongation takes place at the temperature where HfC particles begin to lose their effectiveness in strengthening this alloy. It appears that coarsening of HfC particles also decreases the effectiveness of HfC particles in hindering grain boundary sliding.

The present study clearly shows that W-Re-HfC alloys have not only exceptional high-temperature tensile strength but excellent resistance to high-temperature creep as shown in Fig. 7. The creep strength of W-4.0Re-0.33HfC appears to be better than any high-temperature alloys ever developed. The creep mechanism is often deduced from the activation energy for creep. The activation energy can be determined from the slope

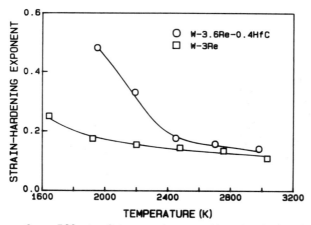

Figure 9 - Effect of temperature on the strain-hardening exponent of W-3.6Re-0.4HfC.

of a line by plotting the steady-state creep rate at a constant applied stress as a function of the reciprocal of the test temperature, as shown in Fig. 10. In the present study, the activation energy for creep of W-4.0Re-0.33HfC is calculated to be about 105 Kcal/mole in the temperature range of 1955 to 2190 K. It appears that the activation energy remains constant within the stress range of 20 to 50 MPa.

The activation energy for creep usually corresponds to an activation energy for bulk diffusion at high temperatures. The activation energies for self-diffusion of tungsten and diffusion of rhenium in tungsten are reported to be 153 and 163 Kcal/mole (15), respectively. The activation energy for creep of W-4.0Re-0.33HfC is found to be 105 Kcal/mole in the present study. Therefore, the creep deformation of W-4.0Re-0.33HfC is not controlled by bulk diffusion. The activation energies for grain boundary diffusion and for dislocation core diffusion in tungsten are reported to be 92 and 90 Kcal/mole, respectively (16). Since the stress exponent is 5.2 for W-4.0Re-0.33HfC, grain boundary diffusion is not considered to be a dominant factor for the creep behavior of this alloy. In the case of creep controlled by grain boundary diffusion, the stress exponent is expected to be about one (17). Therefore, the dislocation core diffusion is considered to be the dominant factor in the creep deformation of W-4.0Re-0.33HfC in the temperature range of 1955 to 2190 K.

Conclusions

The high-temperature mechanical properties of W-3.6Re-0.4HfC were examined with tensile tests above 0.5 T_m. The creep behavior of W-4.0Re-0.33HfC was examined by step-load and step-temperature creep tests. The following conclusions can be drawn from the present study.

(1) The tensile strength of W-3.6Re-0.4HfC is much greater than that of pure W or W-3.6Re above 0.5 T_m. The excellent high-temperature strength of W-3.6Re-0.4HfC is attributed to the strengthening effect of finely dispersed HfC particles.

(2) The contribution of HfC particles to tensile strength decreases significantly above 2700 K. This is considered to be caused by the coarsening of HfC particles.

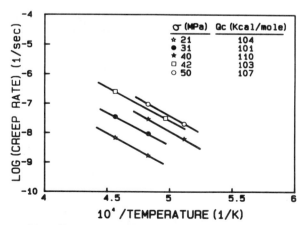

Figure 10 - Temperature dependence of the steady-state creep rate for W-4.0Re-0.33HfC.

(3)   The fracture mode of W-3.6Re-0.4HfC above 0.5 T_m is ductile dimple tearing.   The tendency for ductile tearing increases with temperature.   Extensive grain boundary sliding is responsible for a rapid increase in tensile elongation above 2450 K.

(4)   The W-4.0Re-0.33HfC alloy has excellent high-temperature creep strength.   The stress exponent for the steady-state creep is found to be 5.2 in the temperature range of 1955 to 2783 K.

(5)   The activation energy for creep of W-4.0Re-0.33HfC is found to be 105 Kcal/mole in the temperature range of 1955 to 2190 K. Dislocation core diffusion is considered to be the dominant factor in the creep deformation of W-4.0Re-0.33HfC.

(6)   The creep rupture is initiated by intergranular wedge-shape cracks caused by grain boundary sliding.

Acknowledgements

This work was performed under the Department of Energy contract ACO3-87SF-17170.   The authors would like to thank Mr. S. S. Kim for his kind assistance in temperature measurements.

References

1. W. D. Klopp and P. L. Raffo, "Effects of Purity and Structure on Recrystallization, Grain Growth, Ductility, Tensile, and Creep Properties of Arc-Melted Tungsten" (NASA Tech. Note D-2503, Cleveland, Ohio, 1964).

2. W. D. Klopp, W. R. Witzke, and P. L. Raffo, "Mechanical Properties of Dilute Tungsten-Rhenium Alloys" (NASA Tech. Note D-3483, Cleveland, Ohio, 1966).

3. L. S. Rubenstein, "Effects of Composition and Heat Treatment on High-Temperature Strength of Arc-Melted Tungsten-Hafnium-Carbon Alloys" (NASA Tech. Note D-4379, Cleveland, Ohio, 1968).

4. P. L. Raffo, "Yielding and Fracture in Tungsten and Tungsten-Rhenium Alloys," J. Less-Common Metals, 17 (1969), 133-149.

5. W. D. Klopp and W. R. Witzke, "Mechanical Properties of Arc-Melted Tungsten-Rhenium-Hafnium-Carbon Alloys" (NASA Tech. Note D-5348, Cleveland, Ohio, 1969).

6. W.D. Klopp, P. L. Raffo, and W. R. Witzke, "Strengthening of Molybdenum and Tungsten Alloys with HfC," J. Metals, 6 (1971), 27-38.

7. J. R. Stephens and W. R. Witzke, "Alloy Softening in Group VIA Metals Alloyed with Rhenium," J. Less-Common Metals, 23 (1971), 325-342.

8. W. D. Klopp and W. R. Witzke, "Mechanical Properties of a Tungsten-23.4 Percent Rhenium-0.27 Percent Hafnium-Carbon Alloy," J. Less-Common Metals, 24 (1971), 427-443.

9. W. R. Witzke, "The Effects of Composition on Mechanical Properties of W-4Re-Hf-C Alloys," Metall. Trans., 5 (1974), 499-504.

10. L. B. Lundberg, "Refractory Metals in Space Nuclear Power," _J. Metals_, 37 (4)(1985), 44-47.

11. R. R. Vandervoort, "The Creep Behavior of W-5Re," _Metall. Trans._, 1 (1970), 857-864.

12. R. R. Vandervoort and W. L. Barmore, "Elevated Temperature Deformation and Electron Microscope Studies of Polycrystalline Tungsten and Tungsten-Rhenium Alloys," _Proc. 6th Plansee Seminar_, Metallwerk Plansee Ag., Reutte-tyrol, 1969, 108-137.

13. W. D. Klopp, W. R. Witzke, and P. L. Raffo, "Ductility and Strength of Dilute Tungsten-Rhenium Alloys," _Refractory Metals and Alloys IV-Research and Development_, ed. R. I. Jaffee et al. (New York, NY: Gordon and Breach, 1967), 557-573.

14. J. L. Taylor, "Tensile Properties of Tungsten-3% Rhenium from 1400 to 2900°C in Vacuum," _J. Less-Common Metals_, 7 (1964), 278-287.

15. R. L. Andelin, J. D. Knight, and M. Kahn, "Diffusion of Tungsten and Rhenium Tracers in Tungsten," _Trans. TMS-AIME_, 233 (1965), 19-24.

16. K. G. Kreider and G. Bruggeman, "Grain Boundary Diffusion in Tungsten," _Trans. TMS-AIME_, 239 (1967), 1222-1226.

17. F. Garofalo, _Fundamentals of Creep and Creep-rupture in Metals_ (New York, NY: The MacMillian Co., 1965), 197.

THE EFFECT OF THORIA ON THE TENSILE PROPERTIES
OF SINTERED W-30W/O RE ALLOY AT ELEVATED TEMPERATURES*

B.H. Tsao and D. L. Jacobson

Department of Chemical, Bio and Materials Engineering
Arizona State University
Tempe, AZ   85287

ABSTRACT

An investigation was made of the tensile properties of sintered W-30Re in the temperature range 2000-2750K in order to understand the mechanism by which thoria increases the strength of W-Re alloys. It was found that both the strength and ductility of thoriated material is improved. Based upon metallographic observations, most cracks initiate at grain-boundary triple points, propagate along grain boundaries and form voids roughly normal to the tensile axis in both thoriated and unthoriated specimens. However, the thoriated specimens have fewer voids and a smaller grain size. It was also observed that most of the thoria particles lie at grain boundaries along the tensile axis. SEM micrographs of the fracture surfaces confirmed that the fracture mode of thoriated and unthoriated materials is intergranular. The dislocation-particle interaction theories do not adequately explain the tensile results due to the preferential distribution of thoria particles. In general, grain boundary sliding at high temperatures has been postulated to be the primary factor giving rise to the formation of voids at grain boundaries. The present study suggests that fewer voids and higher twin densities could be the reason why thoriated specimens have higher ductility and that thoria particles at the grain boundaries impede grain boundary sliding, leading to the observed increase in the strength of W-Re alloys.

* Research supported by the Innovative Science and Technology Directorated of the U.S. Department of Defense Strategic Defense initiated through the Air Force Wright Aeronautic Laboratories of the Wright-Patterson Air Force Base.

Refractory Metals: State-of-the-Art 1988
Edited by P. Kumar and R.L. Ammon
The Minerals, Metals & Materials Society, 1989

Introduction

It has been shown that additions of rhenium and thoria increase the strength and ductility of tungsten at temperatures ranging from room temperature to 1873K (1,2,3). Maykuth et al (4) have found a lower ductile-brittle transition temperature in a W-5Re-2.2ThO$_2$ sheet alloy made by powder metallurgy techniques. King et al (5) have produced evidence that solid solution strengthening and dispersed second phase strengthening are additive effects in W-25Re-1ThO$_2$. The present work was undertaken with the objective of understanding the strengthening mechanism of thoria. The influence of 1 w/o thoria on the tensile properties of W-30Re has been investigated here. The results are compared with the tensile properties of W-1ThO$_2$ and W-25Re-1ThO$_2$.

Experimental Procedures

The test materials were fabricated by Rhenium Alloys Inc., Cleveland, Ohio. The tungsten-rhenium-thoria alloys were sintered from tungsten, rhenium and thoria powder. The alloys were fabricated at a pressure of 200 MPa and a temperature of 2500K, then swagged to the desired diameter. The process was carried out in a reducing atmosphere of ultra pure hydrogen. The furnace used for this purpose had a lining of molybdenum to withstand the high temperature.

The tensile testing of the specimens was conducted on an Instron with a high vacuum chamber. The flat specimens described in (6) were first mechanically polished with 0.05μ alumina and chemically polished to ensure the same surface condition. All specimens were heated at 2500K and recrystallized in a vacuum chamber at 10^{-7} torr for an hour to obtain a stable structure. Temperatures were measured with a photon counting pyrometer calibrated with an NBS ribbon filament lamp (7,8), then corrected by using the normal spectra emissivity data of W-30Re-1ThO$_2$(9) and W-30Re(10). All experiments were conducted at the strain rate of 7 x 10^{-4} sec^{-1}. The fractured specimens were examined by both optical and scanning electron microscopes.

Experimental Results

The test results (average of at least two samples) have been summarized in Fig. 1 to Fig. 4. Ultimate tensile strength (U.T.S.), 0.2% offset yield strength, reduction of area and normalized Young's modulus are plotted as a function of testing temperature. The results (11) of W-25Re-1ThO$_2$ and W-1ThO$_2$ are also plotted for comparison as shown in Fig. 1 and Fig. 2. The ultimate tensile strength of thoriated alloys was approximately 21 MPa higher than that of unthoriated alloys. Also the 0.2% offset yield strength of thoriated specimens was approximately 24 MPa higher than that of unthoriated alloys.

Fig. 3 shows that the normalized Young's modulus of W-30Re alloys with respect to Young's modulus of pure tungsten at room temperature decreases with increasing testing temperature. This trend agrees well with previous results (12).

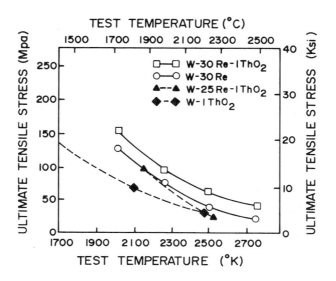

Figure 1 - Temperature dependence of ultimate strength of W-Re
and W-Re-1ThO$_2$ alloys.

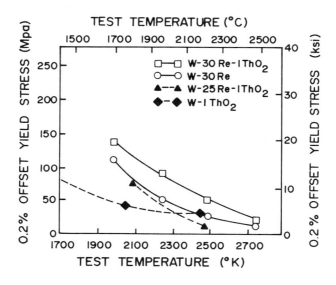

Figure 2 - Temperature dependence of 0.2% Yield strength of W-Re
and W-Re-1ThO$_2$ alloys.

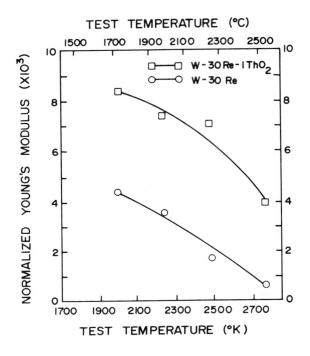

Figure 3 - Temperature dependence of normalized Young's modulus
of W-30Re and W-30Re-1ThO$_2$ alloys.

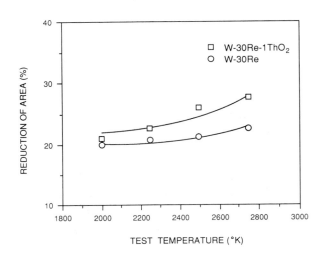

Figure 4 - Temperature dependence of ductility of W-30Re and W-
30Re-1ThO$_2$ alloys.

Figure 4 shows that the percentage reduction of area increases with increasing test temperature. However, the percentage of reduction of area of W-30Re-1ThO$_2$ is always higher than that of W-30Re within the exper-experimental regimes.

Figure 5 shows a typical fractograph of an unthoriated specimen at temperature 2750K which exhibits intergranular fracture. The typical grain size of the unthoriated material is about 52μm.

Figure 6 shows a typical fractograph of a thoriated material. The fracture mode is also intergranular. The unthoriated grain size is approximately 8μm (estimated by the intercept method) larger than that of thoriated material.

Figure 7(a) is a typical micrograph of the side of the fractured unthoriated W-Re specimens near the fractured end. It is seen that voids form at grain boundaries and are roughly normal to the tensile axis (horizontal direction). Fig. 7(b) shows the same specimen as Fig. 7(a), but slightly away from the fractured end. It is seen that voids form at grain boundaries and twin structures form within grains with a random distribution. Fig. 7(c) shows the same specimen as Fig. 7(a) at higher magnification, but away from the fractured end. It is seen that most voids form at grain boundary triple points. Occasionally, it is found that voids form at parallel grain boundaries. However, the concentration of voids formed at parallel grain boundaries is often much less than voids formed at triple grain boundary points. It is likely that most cracks initiate at grain boundary triple points.

Fig. 8(a) is a typical micrograph of the side of the fractured thoriated sample near the fracture end. The voids form at grain boundaries, roughly normal to the tensile axis. This is essentially the same for the unthoriated specimens. An additional feature is that most thoria particles lie at grain boundaries along tensile axes. Fig. 8(b) shows the same specimen as Fig. 8(a), but slightly away from the fractured end. It is seen that twin structures form within grains such as in the unthoriated specimens. However, the twin density of thoriated specimens is higher than that of the unthoriated specimens. The typical value of the twin density of the thoriated specimen at 2750K is 5.42 x 10^{-2} mm/mm^2 where as the unthoriated value is 4.41 x 10^{-2} mm/mm^2. Fig. 8(c) shows the same specimen as Fig. 8(a) at higher magnification, but away from the fractured end. It is clearly seen that twin structures form within grains.

Figure 9(a) is a secondary electron image which confirms that most thoria particles lie at grain boundaries. Figure 9(b) shows the double exposured micrograph of backscattered-electron and characteristic x-ray images. It is observed that most thoria particles lie at grain boundaries along tensile axis. Figure 9(c) is a micrograph of backscattered-electron (left) and an x-ray characteristic image (right). It confirmed that thoria particles lie along tensile axes.

100μm

Figure 5 - Fractograph of W-30Re alloy tested at 2750K.

100μm

Figure 6 - Fractograph of W-30Re-1ThO$_2$ alloy tested at 2750K.

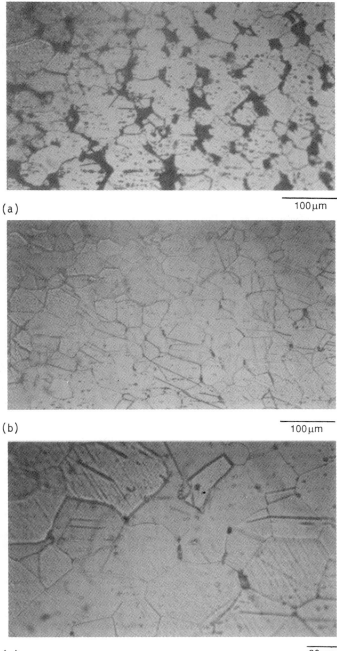

(a)   100μm

(b)   100μm

(c)   20μm
Figure 7 - Micrograph of the side of W-30Re alloys fractured at
2750K: (a) near to, (b) slightly away from (c) away from the
fractured end.

83

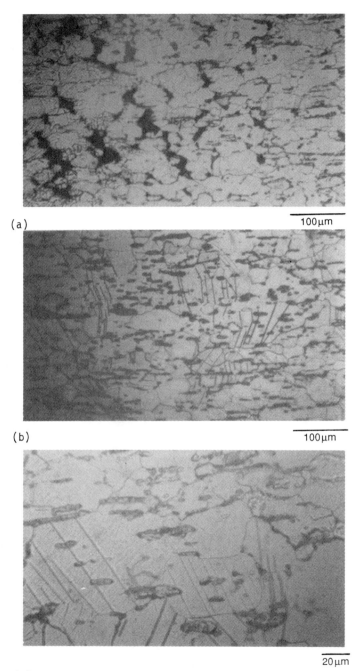

(a)

100µm

(b)

100µm

(c)

20µm

Figure 8 - Micrograph of the side of W-30Re-1ThO$_2$ alloys fractured at 2750K: (a) near to, (b) slightly away from and (c) away from the fractured end.

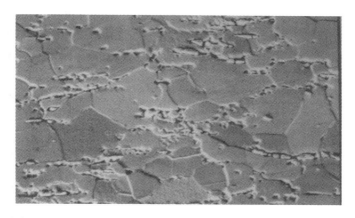

(a)

(b)

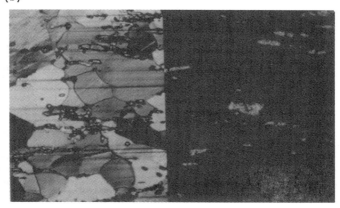

20μm

(c)
Figure 9 - Electron fractographs of W-30Re-1ThO$_2$ fractured at
2750K: (a) secondary-electron (b) backscattered-electron images
(c) backscattered-electron (left) and x-ray (right) images.

85

Discussion

The typical engineering stress-strain curves for W-Re and W-Re-ThO$_2$ shows that work hardening precedes plastic instability. However, the stress-strain curve with no work hardening and a load decrease after 1% deformation reported by King et al (13) was not observed in the present study.

The U.T.S. and 0.2% offset yield strengths both decreased with increasing test temperature as shown in Figure 1 and Figure 2. However, the strength of W-30Re-1ThO$_2$ specimens is substantially higher than that of W-30Re specimens up to about 2473K. The addition of ThO$_2$ particles to W-30Re caused an increase in the ultimate and yield strengths. In comparison, the yield strength of W-30Re-1ThO$_2$ is only slightly higher than that of W-30Re at 2473K. The temperature dependence of the ultimate strength of the alloys is similar to that of the yield strength. The ultimate and yield strengths of each of the W-25Re-1ThO$_2$ and W-1ThO$_2$ alloys are almost the same at temperatures near 2473K. King et al (13) explained that the U.T.S. and yield strengths of W-Re alloys are about the same at temperatures above 1873K, which is due to the absence of work hardening.

It is shown that W-30Re-1ThO$_2$ has better ductility than W-30Re, based upon the percentage of reduction of area. Also from metallographic observations, as shown in Figure 7 and Figure 8, it is determined that the thoriated materials have fewer voids than unthoriated materials. King et al (13) reported that above 1923K, the percent elongation at fracture of W-25Re-1ThO$_2$ is increasing. The presence of twin structures in both W-30Re and W-30Re-1ThO$_2$ constitutes the major difference between that of the present study and the previous work of King et al. They explained that the ductility behavior of W-Re alloys is due to the opposing effect of void formation and dynamic recovery (stress-aided cross slip), i.e., the decrease in ductility being that the result of the former, whereas the increase of ductility results from the latter. In the present case, the existence of twin structures could reduce the activation energy for cross slip and, therefore, dynamic recovery could take place more easily.

In all specimens, the mode of fracture is intergranular. Previous work (16) has shown that voids are formed at grain boundaries in pure tungsten and W-1ThO$_2$ during deformation and that these lead to intergranular fracture. Voids also formed in the uniform gauge section for both sets of specimens but to a lesser extent than was observed in the fracture end. Some voids were present at the parallel boundaries but the majority were in the grain boundary triple points. Grain size was measured in both the uniform gauge section and the unstressed portion. It was found that grain size increased with increasing test temperature in both the gauge section and the unstressed portion. Since the grain size difference between thoriated and unthoriated material is about 8μ, it is concluded that the effect of grain size is minimal.

The difference in yield and ultimate stress between W-Re and W-Re-ThO$_2$ can be interpreted by evaluating two parameters: grain size, particle size and distribution. As mentioned earlier and shown in Fig. 9, due to the small difference of the grain size and the preferential distribution of ThO$_2$ particles, the traditional dislocation-particle interaction theories proposed by Orowan (17) and Fisher et al (18) do not adequately explain the present results. It is also noted that ThO_2 particles are not present at voids, i.e., most grain boundaries which have ThO_2 particles do not form voids. This is evidence that ThO_2 particles in some way impede grain boundary sliding. In general, grain boundary sliding at high temperature has been postulated to be the primary factor giving rise to the formation of voids at grain boundaries. The present study suggests that fewer voids and higher twin densities could be responsible for thoriated specimens having higher ductility and that thoria particles at grain boundaries impede grain boundary sliding, thus increasing the strength of W-Re alloys.

Conclusion

1) The dispersed second phase (1 w/o ThO_2) increase the strength and ductility of W-30Re alloy.
2) The increase in strength is not explained by the dislocation-particle interaction theory due to the size and preferential distribution of ThO_2.
3) Reduced void formation and higher twin densities are responsible for the higher ductility of W-Re-ThO_2 alloy.
4) Improved alloy strength results when thoria particles impede the grain boundary sliding.

Acknowledgement

Authors would like to thank Mr. Rick Stout for his assistance to carry out experiments and Ms. Vicky Wright for preparing manuscript.

References

1. G. A. Geach, and J. E. Hughs, "The Alloys of Rhenium with Molybdenum or with Tungsten Having Good Temperature Properties," Proceedings, Second Plansee Seminar, Metallwork Plansee. A. G. Reutte, Austria, (1955), 246.

2. R. I. Jaffee, C. T. Sims, and J. J. Harwood, "The Effect of Rhenium on the Fabricability and Ductility of Molybdenum and Tungsten," Plansee Proceeding 1958, Pergamnon Press, Ltd, London, (1959) 380.

3. G. T. Hahn, A. R. Rosenfield, "A Modified Double-Pile-Up of the Grain Size and Dispersion Particles of the Influence on Brittle Fracture," Acta Metallurgical, 14 (1966), 1815-1825.

4. J. D. Maykuth, F. C. Holden, and R. I. Jaffee, "The Workability and Mechanical Properties of Tungsten and Molybdenum-Base Alloys Containing Rhenium," (paper presented at the 113th Meeting of Electoro-Chemical Society; Chicago, May 3-4, 1960).

5. G. W. King, "An Investigation of the Yield Strength of a Dispersion-Hardening W-3.8 Vol. Pct ThO_2 alloys," Transactions of the Metallurgical Society of AIME, 245, (1969), 83-89.

6. M. L. Ramalingam, "Investigation of Sintered Tungsten, Rhenium Additive Alloys for High Temperature Space Application," (Ph.D. Dissertation, Arizona State University, May 1985) 135.

7. J. R. D. Lee, "The International Practical Temperature Scale of 1968," Metrologia, 5 (1969), 35-47.

8. H.J. Kostkowski and R.D. Lee, "Theory and Methods of Optical Pyrometry," NBS Monograph 41, (1962) 1-20.

9. N.O. Moraga, "High Temperature Heat Transfer and Thermionics Properties of Tungsten Alloys," (Ph.D. Dissertation, Arizona State University, May 1988) 116-117.

10. N.O. Moraga and D. L. Jacobson, "High Temperature Emissivity Measurements of Tungsten-Rhenium Alloys," (Paper presented at AIAA 22nd Thermophysics Conference; Honolulu, Hawaii, June 8-10 1987), AIAA-87-1584.

11. S. W. H. Yin and C. T. Wang, Tungsten-Source, Metallurgy, Properties and Application, Plenum Press, New York and London, 1979).

12. R. Lowrie and A. M. Gonas, "Dynamic Elastic Properties of Polycrystalline Tungsten, 24-188°C." Journal of Applied Physics, 36,(1965), 2189-2192.

13. R. I. Jaffee et al e.d., Refractory Metal and Alloy IV Res. Dev., Vol. I (Gordon and Breach, New York 1969), 621, King et. al.

14. R. W. Cahn ed., <u>Physical Metallurgy</u> (North Holland, London, 1970) 948, J. Weertman and J. R. Weertmann, 1032, P. Hansen.

15. D.L. Davidson and F.R. Brotzen, "Plastic Deformation of Molybdenum-Rhenium Alloy Crystal," <u>Acta Metallurgical</u>, 18, (1970) 463-470.

16. G.W. King and H.G. Sell, "The Effect of Thoria on the Elevated-Temperature Tensile Properties of Recrystallized High Purity Tungsten," <u>Transactions of the Metallurgical Society of AIME</u>, 223, (1965), 1104-1113.

17. E. Growan, <u>Symposium on Internal Stress in Metals and Alloys, Inst. of Metals Monograph and Report Series, No. 5</u>, (Institute of Metals, London 1948), 451.

18. J. C. Fisher, E. W. Hart and R. H. Pry., "The Hardening of Metal Crystals by Precipitate Particles," <u>Acta Metallurgical</u>, 1 (1953) 336-339.

CORROSION OF REFRACTORY METALS AND ALLOYS

IN MOLTEN LiF AT 1173 K

Y. Desai[1], K. Vedula and A. K. Misra[2]

Graduate Student, Associate Professor and
Senior Research Associate, respectively
Department of Materials Science and Engineering
Case Western Reserve University, Cleveland, Ohio 44106

Abstract

LiF salt is being considered as latent heat of fusion phase change material for thermal storage in advanced solar dynamic space power systems. In this investigation corrosion studies have been conducted in molten LiF at 1173 K to determine the suitability of several refractory metals and alloys as containment materials. The refractory metals and alloys studied are Mo, W, Ta, Nb, Nb-lat% Zr and Nb-5a%Zr. Mo and W are quite corrosion resistant in molten LiF, whereas Ta undergoes large extent of corrosion in molten LiF. Nb has moderate corrosion resistance. The rate of corrosion of Nb-Zr alloys is higher than that of pure Nb, the extent of corrosion increasing with increase in Zr concentration of the alloy.

1. Currently with Presrite Corporation, 3665 E. 78th St., Cleveland, Ohio 44105.

2. Currently with Sverdrup Technology, Inc., NASA Lewis Research Center Group, 21000 Brookpark Road, Cleveland, Ohio 44135.

Refractory Metals: State-of-the-Art 1988
Edited by P. Kumar and R.L. Ammon
The Minerals, Metals & Materials Society, 1989

Introduction

Advanced solar dynamic systems are being considered for space power applications [1]. These systems require thermal energy storage units to store excess energy during the sunlit portion of each orbit and to provide thermal energy during the eclipse portion of the orbit. Latent heat storage systems based on the solid to liquid phase transformation are the most attractive means for thermal storage. One of the prime requirements for a latent heat of phase change material for space power applications is a high heat of fusion per unit mass and per unit volume. Furthermore, the melting point of the phase change material must be compatible with the operating temperature of the engine. Although higher operating temperatures are desired for better engine efficiency, lack of high temperature structural materials for temperatures beyond 1273 K limits the operating temperatures for the heat engines. Lithium fluoride, because of its high energy density (1.08 kJ/g and 1.93 kJ/cm^3) and a reasonably high melting point (1121 K), is the prime candidate as a phase change material for space power applications.

The container materials to be used for these systems must, therefore, be resistant to corrosion by the fluoride salt in addition to having the required strength at elevated temperature. Although corrosion of metals and alloys in fluoride salt melts have been studied in some detail for fluoride melts with melting points lower than 973 K, corrosion data in fluoride melts at higher temperatures are rather limited and mostly restricted to NaF-based eutectic systems [2].

Refractory metals are likely candidates for thermal energy storage container materials because of their higher strength at elevated temperatures and, hence, the focus of this investigation has been to determine the compatability of several refractory metals and alloys with LiF salt melts and to obtain an insight into the mechanisms of material degradation in LiF melts.

Thermodynamic Considerations

The corrosion of a metal M in pure LiF melt can occur according to the reaction:

$$M(s) + xLiF(l) = MF_x + Li(l) \qquad (1)$$

The relevant metal fluorides of interest in the present study have low boiling points as can be seen from Table I. Therefore, the reaction of Mo, W, Ta, Nb, and Zr with pure LiF melt would lead to formation of gaseous metal fluoride species. The pertinent reactions along with their Gibbs free energy changes ($\Delta G°$) at 1173 K are given below [3]:

$$Nb(s) + 5LiF(l) = NbF_5(g) + 5Li(l) \quad \Delta G° = 242.39 \text{ kCal} \qquad (2)$$

$$W(s) + 6LiF(l) = WF_6(g) + 6Li(l) \quad \Delta G° = 389.35 \text{ kCal} \qquad (3)$$

$$Ta(s) + 5LiF(l) = TaF_5(g) + 5Li(l) \quad \Delta G° = 216.75 \text{ kCal} \qquad (4)$$

$$Mo(s) + 6LiF(l) = MoF_6(g) + 6Li(l) \quad \Delta G° = 424.94 \text{ kCal} \qquad (5)$$

$$Zr(s) + 4LiF(l) = ZrF_4(g) + 4Li(l) \quad \Delta G° = 113.96 \text{ kCal} \qquad (6)$$

All the above reactions have large positive ΔG° values. Therefore, the equilibrium partial pressures of the metal fluoride gases will be extremely low. These values, assuming unit activity for Li, LiF and the metal, are given in Table II. Because of these extremely low values for the equilibrium partial pressures for the metal fluoride gases, no reaction would be expected between the metals, W, Mo, Ta, Nb, and Zr with pure LiF melt in a closed system. Some reaction can, however, be expected between the metals and pure LiF melt in an open system where the gaseous reaction products are continuously removed from the system. However, because of very low values for the equilibrium partial pressures of metal fluorides, the rates of these reactions would be extremely slow. Indeed, for all practical purposes, the metals, W, Mo, Ta, Nb, and Zr can be considered to be inert in a pure LiF melt.

LiF is, however, very hygroscopic and, therefore, moisture is a common impurity in LiF salts. Removal of moisture from LiF salts is extremely difficult (3) and the presence of moisture in the salt melt can affect the corrosion behavior of a metal. Reaction of moisture with LiF melt generates HF gas via the reactions:

$$LiF(1) + H_2O(g) = \underline{LiOH}(1) + HF(g) \tag{7}$$

$$2LiF(1) + H_2O(g) = \underline{Li_2O}(1) + 2HF(g) \tag{8}$$

The underline in the above reaction indicates that the species are dissolved in the melt. Some of the HF gas can also dissolve in the melt according to the reaction:

$$HF(g) = \underline{HF}(1) \tag{9}$$

Besides the HF generated by reactions (7) and (8), LiF salt also contains some HF as an impurity.

When a metal sample is inside the LiF melt, the metal would react with LiOH and Li_2O components of the melt to form an oxide and with dissolved HF to form a metal fluoride. Corrosion of a metal in a LiF melt would be a function of the activities of Li_2O, LiOH, and dissolved HF in the melt which in turn are governed by the partial pressures of H_2O and HF above the melt. In an open system, unless the partial pressures of HF and H_2O are fixed above the melt, the activities of Li_2O, LiOH, and dissolved HF in the melt continuously change with time. Therefore, equilibrium calculations for corrosion of a metal in the LiF melt cannot be made in an open system without maintaining the partial pressures of HF and H_2O in the gas phase constant. However, equilibrium calculations can be made for a closed system from a knowledge of the initial moisture and HF content of the salt.

Figure 1 gives a schematic of the closed container with salt in it. The container is only partially filled with the salt. For our calculations we assume the ratio of the open volume in the container to be equal to 100 times the molar volume of the salt melt. The equilibrium compositions of the gas phase, of the melt, and of the solid phase were calculated as a function of the initial moisture and HF content of the salt. These calculations were performed by a computer program called SOLGAS originally developed by Erikson et. al. (4) and are based on the principle of minimization of the total free energy of the system. The gas phase is assumed to consist of HF, H_2O, O_2, H_2, and the metal fluoride gas. The liquid phase is assumed to consist of LiF, Li_2O, LiOH, Li, and dissolved HF. The solid phase is assumed to consist of the metal and its oxide. All the equilibrium calculations were performed at 1173 K.

Although LiOH, Li$_2$O, and the dissolved HF are the actual species that react with the metal, the overall reactions can be considered to be between the metal and H$_2$O + HF. Then, any reaction of metal with H$_2$O and the HF would effectively decrease the partial pressures of these two gases inside the container. Thus a comparison of the partial pressures of HF and H$_2$O inside the container for a metal-LiF-H$_2$O-HF combination with that of a situation in which there is no metal inside the metal would give an indication of the reactivity of the melt for the metal. It is assumed that the container material is inert in the salt melt.

The equilibrium partial pressures of HF and H$_2$O inside the container as a function of initial moisture content of the salt for different metal-LiF combinations are shown in Figures 2 and 3, respectively. Also shown in Figures 2 and 3 are the equilibrium partial pressures of HF and H$_2$O corresponding to LiF only. The partial pressures of HF and H$_2$O for Mo-LiF and W-LiF combinations are slightly lower than those for LiF only. This indicates that Mo and W would be relatively inert in LiF melt even in the presence of moisture. On the other hand, partial pressures of HF and H$_2$O inside the container are significantly decreased by the presence of Ta, Nb, and Zr in the melt, the magnitude of decrease being the maximum for Zr. This suggests that Nb, Ta, and Zr would react with the LiF melt in the presence of moisture and the extent of reaction for Zr is likely to be significantly higher than for the other metals. Presence of Zr in the melt reduces the partial pressure of H$_2$O in the container by about 10 orders of magnitude and that of HF by about 4-5 orders of magnitude. This suggests that, although both ZrO$_2$ and ZrF$_4$ are likely to be formed as a result of reactions of Zr with H$_2$O and HF, the amount of ZrO$_2$ formed would be higher than that of ZrF$_4$. In the presence of moisture, the reactivity of melt towards both Nb and Ta are similar. Again, in the case of Nb and Ta, the decrease in partial pressure of H$_2$O is much greater than the decrease in partial pressure of HF, suggesting that the amount of metal oxides formed could be higher than that of metal for Nb and Ta as well.

The effect of different metals on the equilibrium partial pressures of HF and H$_2$O inside the container for LiF salts containing 100 ppm moisture as well as 1000 ppm HF as impurities is shown in Figure 4. The partial pressures of HF and H$_2$O inside the container for melts containing W and Mo are nearly the same as that for LiF only. Thus both W and Mo would be relatively inert in LiF melts even if HF is present in the salt as an impurity in addition to moisture. The metal Zr would suffer from maximum extent of reaction as evidenced by large decreases in both HF and H$_2$O partial pressures inside the container. The decrease in the partial pressure of H$_2$O inside the container is nearly the same for both Nb and Ta. However, the decrease in the partial pressure of HF inside the container is slightly greater for Ta as compared to that of Nb. Thus, the extent of reaction of Ta in a salt melt containing both HF and H$_2$O as impurities would be slightly greater than that of Nb.

In summary, W and Mo would be relatively inert in the salt melt even if moisture and HF are present as impurities. The metals Ta, Nb, and Zr would react with the salt melt when either moisture alone or both moisture and HF are present as impurities. In the presence of impurities like H$_2$O and HF, the reactivity of the melt towards the metals decreases in the following order: Zr, Ta, Nb. The experimental investigation described below was undertaken to verify some of these thermodynamic predictions and to examine the kinetics and mechanisms of attack.

Materials and Procedure

Pure refractory metals Mo, W, Ti, Ta and Nb were obtained from Materials Research Corporation and a sample of Nb-1at%Zr alloy was obtained from Wah Chung Corporation. A sample of Nb-5at%Zr alloy was prepared by vacuum arc melting, followed by vacuum treatment at 1200°C for 24 hours, at NASA Lewis Research Center, Cleveland. Samples of the alloys were cut into small discs, 12 mm in diameter and .35 mm in thickness and exposed to the salt environment. Samples of the pure metals were in the form of coupons 10 mm in length, 10 mm width and 1.5 mm thick. All specimens were polished to 600 microns using SiC paper before exposure to Lif.

Two different purities of reagent grade LiF salt (regular purity of 99.9% and high purity of 99.99%) were used. The regular purity salt was in the form of solid granules from Cerac Inc. and the high purity salt was in powder form and obtained from Johnson Matthey Materials Tech., U.K. The compositions of the salts are given in Table III. For some experiments, the salt was premelted under argon atmosphere at 1173 K for 24 hours prior to exposing the metal specimen. LiF salt was found to have an inherent moisture content of at least 0.21 to 0.23 wt% and this heat treatment is believed to remove most of the moisture. The gas used in the experiments was pure and ultra high purity argon from Matheson and Liquid Carbonic.

The experimental set-up used is illustrated schematically in Figure 5. The metal specimen and the salt were placed in a graphite crucible since the salt melts do not react with the graphite. The crucible was placed in a quartz tube and evacuated to a pressure of 10^{-2} torr using a mechanical pump. After evacuation, an inert gas atmosphere was maintained throughout the course of the experiment by passing argon gas at a steady rate of 30 cc/min. A zirconium foil was placed around the crucible to serve as an oxygen getter. The quartz tube was then placed in a resistance heated electric furnace. All the specimens were exposed to a temperature of 1173 + 50 K for different periods of time.

After the heat treatment, the specimens were furnace cooled to room temperature prior to removal from the solidified salt. The samples were analyzed by metallography as well as SEM and EDS techniques using a JEOL 840.

Experimental Results and Discussion

The results of the corrosion of the different metals and alloys are presented below and these are discussed in the light of the thermodynamic analysis presented above.

A. Corrosion of Pure Metals

The relative severity of attack observed for pure metals is tabulated in Table IV. Mo and W are the most resistant to attack as illustrated by the cross section of exposed sample of Mo exposed for 24 hours to non-premelted LiF in Figure 6a. Zr was the worst as illustrated by the cross section of a sample exposed to non-premelted LiF for one hour in Figure 6b. Nb shows an uneven attack intermediate between the Mo and the Zr samples as illustrated in Figure 6c. The surface shows uneven and localized penetration into the metal. The reasons for these localized attacks are not clear. The extent of corrosion for Ta is higher than that of Nb. It appears that an oxide layer is present on Ta after corrosion in LiF melt.

The extent of corrosion of pure metals increases in the following order: Mo, W, Nb, Ta, Zr. The corrosive attack of pure metals is believed to be due to the reaction of the metal with H_2O and HF in the melt and the severity of attack is consistent with the predictions based on the above theremodynamic analysis.

B. Corrosion of Alloys of Zirconium in Niobium

The corrosive attack on pure Nb is uneven along the surface and the depth of attack is above 40 microns when exposed for 48 hours to non-premelted LiF. Addition of Zr to Nb intensifies the attack as discussed below.

(a) Nb-1Zr Alloy. This alloy was studied in detail because of its commercial importance. Various specimens were exposed for different time periods ranging from 1 hour to 250 hours in both non-premelted and premelted LiF.

Figure 7a shows the cross section of a Nb-1Zr alloy specimen exposed for 1 hour in non-premelted LiF. The salient features of the cross section are: (1) a distinct layer of corrosion product completely detached from the alloy surface, and (2) fine precipitates inside the alloy (not readily evident from Figure 7a). Also, uneven corrosion and localized penetration into the alloy are readily evident. X-ray diffraction did not reveal any oxide layer on the surface. Energy dispersive analysis did not reveal any oxygen in the detached layer; only Nb and small amount of Zr was detected in this layer. Thus the detached layer is most likely a layer of metal detached from the alloy surface. It was somewhat difficult to analyze the internal precipitates because of their extremely small size. However, as will be seen later in this section, corrosion of Nb-1Zr in LiF melt for longer times leads to formation of larger precipitates which were identified to be ZrO_2. Thus it is believed that the fine internal precipitates in Fig. 7a are ZrO_2 particles.

The cross section of a Nb-1Zr alloy corroded in non-premelted LiF melt for 70 hrs (illustrated in Figure 7b) shows the same basic features as that of Figure 7a, i.e., (a) detached layer of metal in the melt, and (b) an internal oxidation zone inside the alloy. In addition to a detached metal layer deep inside the melt, there are also some chunks of metal inside the melt at the melt-alloy interface. X-ray maps clearly do not show any oxygen in the detached layers inside the melt. The internal oxide particles inside the alloy are much larger than those seen after 1 hr of corrosion. Also, the internal oxides appear to be interconnected and this gives the distinct appearance of a subscale inside the alloy. X-ray maps show that the subscale is ZrO_2. A thin fluorine-rich layer inside the alloy and the absence of Nb and Zr in this layer probably suggests melt penetration inside the alloy.

The effects of premelting the salt on corrosion morphology for Nb-1Zr alloy are shown in Figures 8a and 8b. The scale morphology after 2 hrs of corrosion (Figure 8a), shows a thin detached metal layer at the melt-alloy interface and an internal subscale. The scale morphology after 250 hours of corrosion (Figure 8b) shows a similar morphology. An added feature of scale morphology after 250 hrs of corrosion is localized penetration (probably of the salt melt) into the alloy at the alloy-melt interface.

Although the features of scale morphology are similar for both pre-melted and nonpremelted salts, the extent of attack is higher for non-premelted salts. This is illustrated in Figure 9 which shows the depth of attack as a function of time of exposure for both premelted and non-premelted salts. This is to be expected since premelting the salt removes some moisture and HF from the salt. However, the results from this study demonstrate that premelting of the salt does not completely remove the moisture and HF from the salt.

(b) Corrosion of Nb-5Zr. With a larger amount of Zr in the alloy, the attack is much more severe, although the general features of the corro-sion morphology are similar to that of Nb-1Zr alloy. Figure 10 illustrates the corrosion morphology for this alloy exposed in premelted LiF for 48 hours. Large chunks of metals are readily evident inside the melt.

(c) Corrosion Mechanisms for Nb-Zr Alloys. Corrosion of pure Nb results in localized penetration of melt inside the metal. This is also observed for Nb-Zr alloys. However, in addition to localized melt pene-tration, corrosion morphology for Nb-Zr alloys shows detachment of layers of metal from the alloy surface. There is no indication of formation of any niobium oxide for corrosion of either pure Nb or Nb-Zr alloys. On the other hand, a subscale of ZrO_2 is formed for Nb-Zr alloys. This suggests that the oxygen potential inside the melt is lower than that corresponding to Nb-NbO equilibria, but it is still greater than that corresponding to Zr-ZrO_2 equilibria. It seems likely that localized penetration of the melt into the alloy coupled with the formation of internal ZrO_2 particles is responsible for detachment of metal layers from the alloy surface, and this is explained as follows.

Initially the corrosion of Nb-Zr alloys, similar to that of pure Nb, is due to localized melt penetration. However, since the oxygen potential inside the melt is high enough, formation of ZrO_2 occurs. Because of low Zr concentration in the alloy, formation of a continuous layer of ZrO_2 is not possible. Instead, ZrO_2 is formed as internal oxide particles.

Formation of oxides in LiF melts is due to reaction of metal with LiOH and Li_2O components of the melt. Any decrease in these activities would lead to increase in the activity of dissolved HF in the melt at the melt-alloy interface. Since the oxygen potential in the melt is not sufficient for NbO formation, reaction of dissolved HF with Nb will take place leading to the formation of $NbF_5(g)$ and $H_2(g)$ via:

$$Nb + 5HF + NbF_5 + 5/2H_2 \qquad\qquad (10)$$

Since NbF_5 and H_2 are both gaseous products, the layer of metal that is already penetrated by the melt at localzied areas can be detached from the alloy if the total pressure of NbF_5 and H_2 at the melt-alloy interface in these localized spots (the melt-alloy interface is already inside the alloy because of localized melt penetration) becomes greater than 1 atm. Thermodynamic calculations show that the total pressure of NbF_5 and H_2 at the localized melt-alloy interface can be greater than 1 atm if the pHF at this interface is higher than 0.078 atm. It appears that such values for pHF are possible at the localized melt-alloy interfaces if oxygen can be removed from the melt by oxide formation.

Two key factors appear to be responsible for detachment of metal layers from the alloy surface for Nb-Zr alloys. These are: (1) localized penetration of melt inside the alloy, and (2) removal of oxygen from the melt by oxide formation. For pure Nb, although localized penetration of melt occurs, metal detachment from the surface does not occur because of absence of oxide formation. For Nb-Zr alloys both the conditions are satisfied and detachment of the metal from the alloy surface occurs. A continuous oxide layer is not formed for Nb-Zr alloys used in this study, and this is probably the reason why layers of metal, instead of an oxide layer, are detached from the alloy surface. If a continuous oxide layer can be formed beneath the melt, as is probably the case for pure Zr, HF(g) can penetrate the scale and metal fluorides plus H_2 would be formed at the scale-metal interface. The situation then becomes similar to that of hot corrosion of metals in sulfate melts (5). If the total pressure of the metal fluoride gas plus hydrogen becomes greater than 1 atm., then rupture of the scale would take place and the scale would be detached from the alloy surface. This is probably the corrosion mechanism for pure Zr.

(d) Corrosion Prevention in Nb-Zr Alloys. Since corrosion in LiF melts is due to the presence of impurities like HF and H_2O, removal of these impurities would decrease the extent of corrosion. It has already been shown earlier that premelting of the salt reduces the extent of corrosion. However, it appears that premelting does not remove these impurities completely. A possible method by which HF and H_2O can be removed from the melt is by adding a reactive metal to the melt. Clearly, Zr would be a leading candidate for this since Zr suffers the maximum extent of corrosion in LiF melts. Figure 11 shows the corrosion morphology for a Nb-1Zr alloy corroded in a mixture of premelted LiF and Zr for 70 hours. The sample did not undergo any attack as was expected because the Zr in the melt removes the impurities like HF and H_2O from the melt. This suggests that corrosion of Nb-Zr alloys in LiF melts can be completely prevented by adding small amounts of Zr to the melt. Further experiments are needed to determine the optimum amount of Zr to be added to the melt for corrosion prevention.

Conclusions

The corrosive behavior of several refractory metals and alloys in LiF melts at 1173 K is summarized here.

a. Among the pure metals, the extent of corrosion increases in the order Mo, W, Nb, Ta and Zr. The corrosion of these metals is due to the reaction of the metal with H_2O and HF in the melt. The attack of pure metals was characterized in some cases by local penetration of the melt into the metal.

b. The Nb-Zr alloys corrode more rapidly than pure Nb. Local penetration of melt coupled with the formation of an internal subscale of ZrO_2 is believed to be responsible for the accelerated attack.

c. Removal of moisture from the melt or addition of Zr to the melt is found to reduce the severity of attack in the Nb-Zr alloys.

References

1. M.O.Dustin et. al., "Advanced Solar Dynamic Space Phase Systems Perspectives, Requirements and Technology Needs," Solar Engg.-1987: Proc. of ASME-JSME Solar Energy Conf., p. 574-88.

2. MSR Prog. Semiann. Prog. Report, July 31, 1960, ORNL-3014, p. 55-58.

3. H.A.Laitinen, W.S.Ferguson, R.A.Osteryoung, "Preparation of Pure Fused Lithium Chloride-Potassium Chloride Eutectic Solvent," J. of Electrochem. Soc., 104(8), August 1957, p. 516-20.

4. G.Eriksson, "Thermodynamic Studies of High Temperature Equilibria," Acta Chemica Scandinavica 25, 1971, p. 2651-58.

5. J.A.Goebel, F.S.Petit, "Na_2SO_4-Induced Accelerated Oxidation (Hot Corrosion) of Nickel," Met. Trans., Vol. 1, July 1970, p. 1943-54.

Table I. Boiling Points for Metal Fluorides

Metal Fluoride	Boiling Point (Degree K)
NbF_5	506
TaF_5	502
WF_6	290
ZrF_4	1179
MoF_6	307

Table II. Equilibrium Partial Pressures for Metal Fluorides for the Reaction $M + xLiF = MF_g + xLi$ at 1173 K (Activities of M, LiF, and Li are assumed unity)

Metal Fluoride	Eq. Part. Press. (atm.)
NbF_5	$6.83*10^{-46}$
WF_6	$2.82*10^{-73}$
TaF_5	$4.09*10^{-41}$
MoF_6	$6.60*10^{-80}$
ZrF_4	$5.82*10^{-22}$

Table III. Composition of Reagent Grade Salts as Reported by the Chemical Analysis

Salt	Chemistry
Regular LiF (99.9% purity)	Li 28.3 wt.%; Ca 260 ppm: Na 110 ppm; Fe 600 ppm
Pure LiF (99.99% purity)	Li 27.9 wt.%; Ca <50 ppm; Na <50 ppm; Fe 60 ppm

---

Table IV. Summary of Microstructural Attack of Various Refractory
Metals and Alloys Caused by Exposure to Molten LiF

Salt	Non-premelted	Premelted
Mo	A	A
W	A	A
Nb	B	B+
Nb-1Zr	B	B+
Nb-5Zr	C	C+
Ta	C	C+
Ti	D	D
Zr	D	D

A - No attack   B - Moderate attack
C - Prominent attack   D - Extensive attack

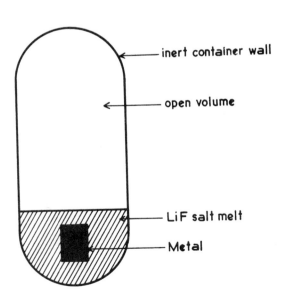

Figure 1.   Schematic of a closed container with salt and specimen.

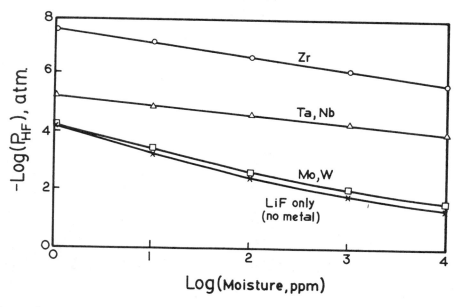

Figure 2. Equilibrium pHF inside the container as a function of initial
moisture content.

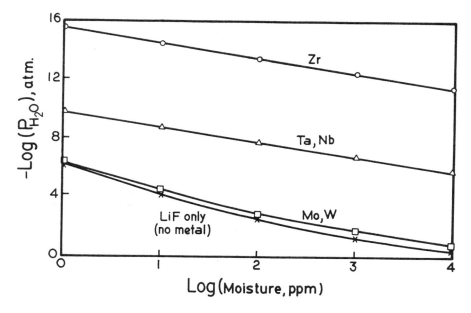

Figure 3. Equilibrium pH_2O inside the container as a function of initial
moisture content.

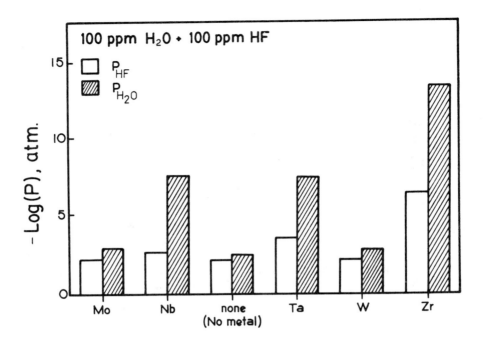

Figure 4.   Equilibrium pH_2O and pHF inside the container for LiF salts
containing 100 ppm H_2O as well as 1000 ppm HF as impurities.

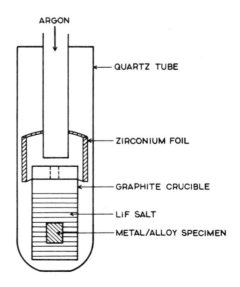

Figure 5.   Schematic of the experimental set-up used.

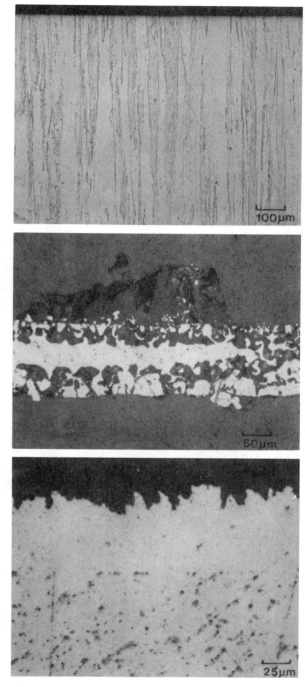

Figure 6. Cross sections of samples of pure metals exposed for 24
hours to non-premelted LiF: (a) Mo, (b) Zr and (c) Nb.

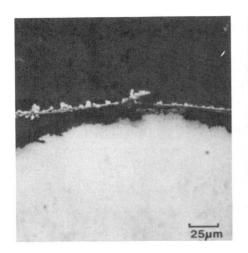 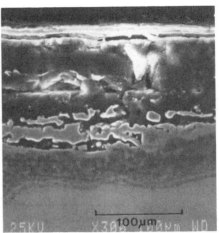

Figure 7. Cross section of sample of Nb-1Zr alloy exposed in non-premelted LiF: (a) for 1 hour, (b) for 70 hours.

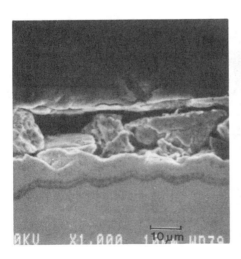 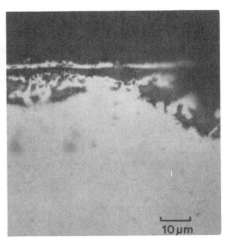

Figure 8. Cross section of sample of Nb-1Zr alloy exposed in premelted LiF: (a) for 2 hours (b) for 250 hours.

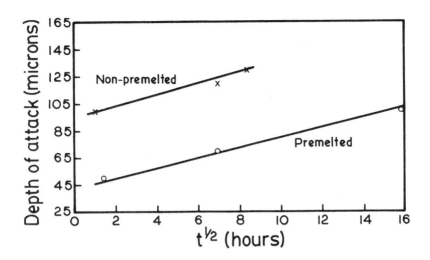

Figure 9. Comparison of depth of attack for Nb-1Zr alloy in premelted and non-premelted salts.

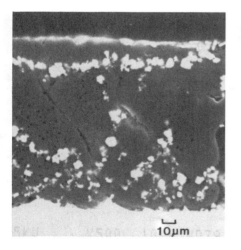

Figure 10. Corrosion of sample of Nb-5Zr alloys exposed for 48 hours to premelted LiF.

Figure 11. Corrosion of sample of Nb-1Zr alloy exposed for 70 hours to premelted LiF containing Zr.

REFRACTORY METAL STRUCTURES PRODUCED

BY LOW PRESSURE PLASMA DEPOSITION

M.R. Jackson, P.A. Siemers, S.F. Rutkowski, and G. Frind

GE Corporate Research & Development Center
P.O. Box 8
Schenectady, NY 12301

Abstract

Using powders as the feedstock to a plasma gun operating in low pressure, thick deposits have been produced using refractory elements and alloys. Structures were characterized by gas content, density and microstructure. Samples exposed to 1400°C were further characterized by measurement of tensile behavior, and for some alloys, rupture behavior. Fractography was performed on selected materials.

Refractory Metals: State-of-the-Art 1988
Edited by P. Kumar and R.L. Ammon
The Minerals, Metals & Materials Society, 1989

Introduction

Refractory metals are known for their strength capabilities at extremely high temperatures, in the regime of 1200-1700°C (2200-3100°F)(1). Interest in these materials has been renewed recently because of the requirements of potential applications ranging from space-based power generation, to components for the Strategic Defense Initiative, to the national aerospace plane and nearer-term aircraft engine systems(2).

In the past, strength has had to be compromised to some extent to insure that the material was fabricable(3). However, the use of RS (rapid solidification) technologies may offer paths to avoid that compromise. For example, efforts have been undertaken to produce RS powders that may increase the fabricability of these very strong materials(4-6).

The RSPD (rapid solidification plasma deposition) process has been shown previously to allow near-net shape fabrication of complex shapes in high strength superalloy(7) and composite materials(8,9). The purpose of the following discussion is to document efforts to extend the RSPD process to refractory metals and alloys. This effort concentrates on materials that are reasonably fabricable by conventional means, since these are the materials which are most readily available now as powder. However, there appears to be no obvious limitation to application of the process to fabricate complex shaped bodies of materials that would be virtually unprocessable by conventional methods.

Procedures

Powders were purchased from a number of vendors in different powder sizes. Elemental powders of Nb, Mo and W, as well as Nb-base alloys (Nb-1Zr, C-103, Cb-1 and WC-3009) were included. All but the Mo powder appeared to have been produced by fracture, and Nb alloys are known to have been produced by hydriding and subsequently fracturing an ingot and dehydriding the fractured powders.

The powders were injected into an RF plasma gun (manufactured by TAFA, Bow, NH) operating at 50kW or more, which was included in a chamber mechanically pumped to reduce pressure of the gases exiting the plasma gun. Deposition was accomplished by repeatedly traversing a steel tube 3.8 cm in diameter and 12.7 cm in length through the plasma while powders were being injected at rates of 4-8 cm^3/min. A tubular deposit with wall thickness of 0.5-0.6 cm was produced in times of the order of 20 minutes.

The deposits were characterized by measurement of oxygen and nitrogen contents compared to the powders, by density determination, and by metallographic and electron microscopic evaluation of the structure.

Cylindrical blanks were electrodischarge machined from the deposits and treated in Ar at 1400°C (2550°F) for 2h, either in an isothermal furnace or in a hot isostatic press at approximately 100 MPa. Button-head samples were machined from the treated blanks to produce test bars with a uniform gage 0.2 cm in diameter. These samples were tested in tension at 25°C to 1200°C (75°F to 2200°F). Elevated temperature tensile tests were performed in a vacuum of 10^{-5} torr. Rupture tests were performed for selected materials at 1150°C in Ar. Fractography was performed by scanning electron microscopy.

Results and Discussion

Deposition

Plasma conditions were adjusted to produce complete melting of the powders, so that conditions were varied according to particle size and alloy melting range. For W, sound deposits were not produced. There was insufficient power to melt the coarsest W powders. The fine powders could be melted, but they predominantly resolidified before reaching the substrate when the substrate was located at a distance where the substrate itself would not melt in the plasma. This problem may be overcome by cooling the substrate. For all other materials, conditions could be found to produce sound deposits.

Gas contents measured on powders and deposits are listed in Table 1. All deposits reported here had densities which were at least 96% of the full density value of the wrought alloys. As can be seen from the table, contamination by oxygen and nitrogen was most severe for the elemental Nb. This is probably not due inherently to the material. The pure Nb deposits were made considerably earlier in time than were the alloy deposits. In the interim, cleanliness practice in the RSPD processing was improved. Also, plasma conditions used for the alloys produced more volatilization of powders during spraying, so that oxygen on the powder surface may have been removed by that volatilization. For the alloys, both oxygen and nitrogen in the deposits were extremely close to the values measured in the powder sample, with gas content reduction observed in some cases.

108

Table 1
Gas Contamination

	Gases in Powder (ppm)		Gases in Deposit (ppm)	
Powder	Oxygen	Nitrogen	Oxygen	Nitrogen
-140 +200 mesh Nb	410	50	675	230
-200 +325 mesh Nb	600	55	835	360
-325 mesh Nb	2010	130	1960	365
-200 +325 mesh Nb-1Zr	667	36	509	42
-200 +325 mesh C-103	413	66	388	75
-200 +325 mesh Cb-1	136	124	281	85
-200 +325 mesh WC3009	1470	84	1807	150

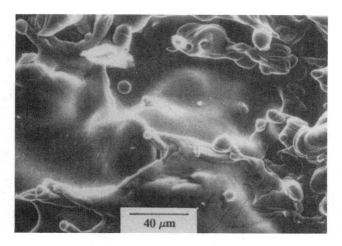

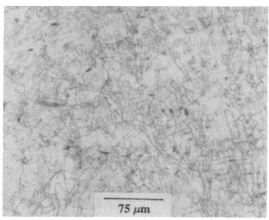

Figure 1   Micrographs of (a) deposit surface (scanning electron microscopy) and (b) cross-section through a deposit of elemental Mo.

The rapid solidification process is characterized by the liquid powder particle droplet impacting and spreading on the substrate surface and solidifying rapidly due to the excellent heat transfer to the substrate. The particle impact and spreading is illustrated in Figure 1, showing the surface of a Mo deposit, and the same deposit in cross-section. A small amount of splashing is evident from the surface view, but the liquid may be quite viscous, based on these micrographs. The splatted particles vary in thickness and extent, depending on the initial particle size and its degree of superheat. For refractory metals made from powders of -200 +325 mesh size (44-74 microns thick), a splat might typically be 14 microns thick and 100 microns in diameter (volume conserved for a 60 micron sphere with no splash).

Similar structures are observed for pure Nb, as seen in Figure 2. Transmission electron microscopy shows the splat interfaces to be marked (at $1/2$ μm intervals in the two dimensional view). It is uncertain whether the marks are voids or small oxide particles. The transmission microscopy shows the splats are multigrained, with grain boundaries running perpendicular to the intersplat boundaries. Some epitaxy occurs from splat to splat, similar to that observed in superalloy deposition(10). The grains are frequently about the same order of magnitude in width as in thickness of the splat, giving a "squared" appearance in the plane perpendicular to the plane of the splats. This is similar to the structures seen in the Mo phase in dc plasma deposited γ/γ'-Mo composite structures(11).

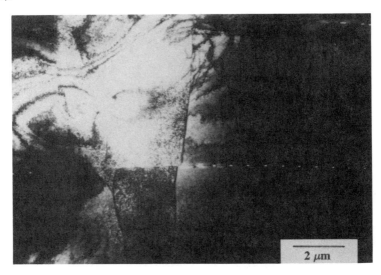

Figure 2   Cross-section transmission electron micrograph through a plasma deposit of elemental Nb.

Heat Treatment

Samples were exposed in argon for 2h at 1400°C (2550°F), either in atmospheric pressure in an isothermal furnace, or in a hot isostatic press at ≈100 MPa. Optically there was no apparent difference between the two exposures. Little grain growth was noted in the alloys or the elemental materials, so that the maximum grain size was of the order of 15 μm, and the average grain size was ≈5 μm. (The elemental deposits were heat treated only, and not HIP'ed.) The layering, evidence of splat boundaries, was far less evident after exposure. In transmission electron microscopy, the grain boundary markers were no longer observed.

Densities were not greatly affected by the exposures. In general, the isothermal exposure increased densities approximately 0.5-1% of theoretical, to average values of 98% of theoretical density. The HIP treatment was more effective, with average values of 98.5% of theoretical density. Presumably, a higher HIP temperature may have resulted in more fully dense structures. Alternatively, the samples could have been sealed in evacuated cans to allow densification of any surface connected porosity. Higher temperatures may cause excessive grain growth, and care must be taken to avoid interaction between the refractory metal and materials it may contact at the elevated temperature.

Mechanical Properties

Tensile tests were performed on heat treated samples for both elemental and alloy deposits, as well as on HIP'ed samples for the alloy deposits. Test results are listed in Table 2. Graphical representations of the data are shown in Figure 3 for 0.2% offset yield strength and elongation at failure.

Table 2
Tensile Behavior of RSPD Refractory Alloys
(all but 25°C tests in vacuum)

Material	Temperature (°C)*	1400C/2h/Ar Strength (MPa) 0.2% Yield	Ultimate	Elongation (%) at Peak Load	at Failure	HIP 100 MPa/1400°C/2h/Ar Strength (MPa) 0.2% Yield	Ultimate	Elongation (%) at Peak Load	at Failure
Nb-1Zr	25	159	253	12	15	173	273	10	11
Nb-1Zr	600	77	208	7	12	101	203	5	6
Nb-1Zr	900	97	186	7	8	107	188	5	7
Nb-1Zr	1200	78	102	5	15	83	110	4	13
C-103	25	282	356	7	9	305	376	6	7
C-103	600	170	279	6	7	161	262	5	6
C-103	900	168	286	6	6	171	275	4	4
C-103	1200	158	165	1	6	147	170	2	8
Cb-1	25	-	502	0	0	-	373	0.1	0.1
Cb-1	600	368	379	0.4	0.5	350	357	0.4	0.4
Cb-1	900	332	356	0.8	0.8	328	339	0.4	0.4
Cb-1	1200	246	256	0.6	1	164	170	0.4	1
WC3009	25	638	662	0.5	0.6	599	629	0.1	0.3
WC3009	600	416	516	3	3	406	454	1	1
WC3009	900	378	383	0.3	0.3	383	441	2	2
WC3009	1200	-	-	-	-	302	321	1	1
Nb 46.5 Ti	25	624	625	0.3	3	707	712	0.2	4
Nb 46.5 Ti	600	112	169	4	14	169	181	2	19
Nb 46.5 Ti	900	64	65	0.3	60	66	66	0.1	54
Nb 46.5 Ti	1200	31	32	0.1	64	-	-	-	-
Nb	25	387	419	2	2				
Nb	560	141	243	7	11				
Nb	710	87	117	6	26				
Nb	860	80	97	9	26				
Nb	1010	66	73	3	33				
Mo	25	-	313	0.1	0.1				
Mo	600	202	285	10	13				
Mo	900	205	270	7	14				
Mo	1200	154	170	4	18				

* 25°C = 76°F
600°C = 1111°F
900°C = 1651°F
1200°C = 2191°F

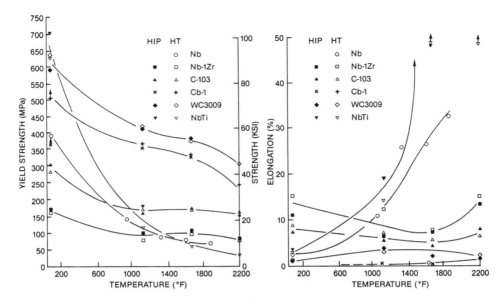

Figure 3   Tensile behavior of plasma deposited refractory metals.

There is no significant difference in behavior between the HIP'ed and heat treated materials: both strength and ductility showed equivalent values for the two exposures. The properties among the different materials rank as expected from data available for wrought alloys, particularly at elevated temperature. The tensile strengths measured for the RSPD materials are in good agreement with values reported for wrought alloys. However, the ductilities are somewhat lower than would be typical of the wrought materials. This is probably due in part to porosity present in the structures, as indicated by measured densities 98-98.5% of the theoretical densities for the thermally treated deposits. The greater oxygen and nitrogen contents in the deposits resulting from high concentrations in the hydrided powders may also be affecting ductility. Although ductilities are reduced compared to wrought materials, all RSPD deposits but Cb-1 have ample ductility for many applications.

Fractography performed on the tensile samples points out some of the features responsible for decreased ductility. Figure 4 shows the fracture surface for the room temperature test of RSPD Mo. Failure is predominantly intergranular, and the columnarity of the grain structure is apparent. Splat boundaries are also evident from the fine porosity seen at those interfaces.

A far more ductile failure is seen in Figure 5 for Nb-1Zr (15% elongation at room temperature). However, there is still evidence of porosity, as well as the presence of a spherodized particle. The powder fed into the plasma is quite angular, due to the fracture process used to produce the powder. The particle shown in Figure 5 apparently melted, but was resolidified prior to incorporation into the deposit.

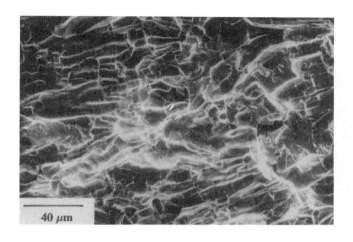

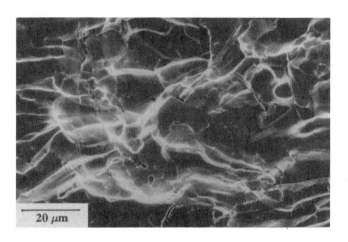

Figure 4   Fracture surface of Mo tested in tension at room temperature.

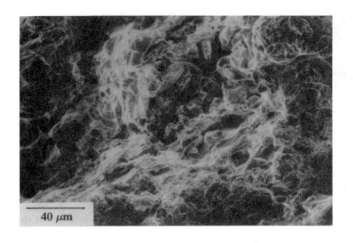

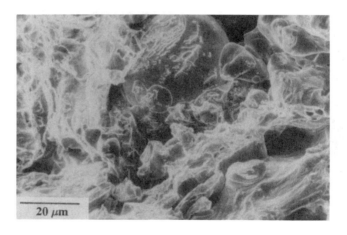

Figure 5   Fracture surface of Nb-1Zr tested in tension at room temperature.

For Cb-1 tested at room temperature (Figure 6), failure was in the elastic regime. The fracture surface indicates a great deal of delamination along splat boundaries, due either to poor bonding of splat to splat or to high concentrations of porosity. Of all the deposits tested, Cb-1 had the lowest density compared to theoretical density. The fracture surface indicates that although melting of the powders was essentially complete, there was little splash of the liquid droplets, and droplet spreading was quite viscous. Evidence of this is seen in the very rounded ends of splats seen in the fracture surface. Empirically, it seems that materials which show splashing of the droplets have relatively good splat-to-splat bonding, and therefore, good ductility. Alloys showing little splash tend to form structures which appear to be poorly bonded, like a stack of pancakes. Hotter plasma conditions may be useful to reduce viscosity of the Cb-1 droplets. This alloy has the highest melting point of the Nb-base materials investigated, although it is still lower than that for Mo.

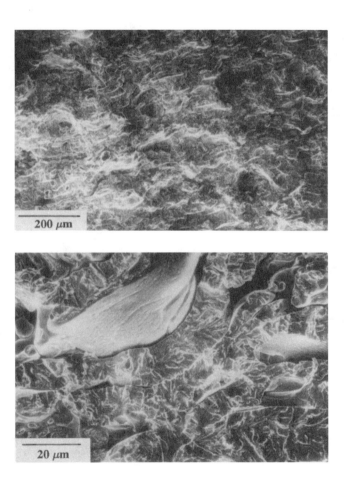

Figure 6   Fracture surface of Cb-1 tested in tension at room temperature.

Rupture tests have been performed on selected alloys after 1400°C (2550°F)isothermal heat treatment. Results are quite similar to those for their wrought counterparts. For example, RSPD Mo lasted 10.5h at 1150°C (2100°F)/83 MPa (12 ksi). This is better than conventionally processed Mo(12). The RSPD sample showed 23% elongation. The comparison is shown in Figure 7. Similarly, RSPD Nb-1Zr lasted 135.4h at 1150°C (2100°F)/69 MPa (10 ksi). This is better than conventionally processed Nb-1Zr(12), as seen in Figure 8. The RSPD sample showed 9% elongation.

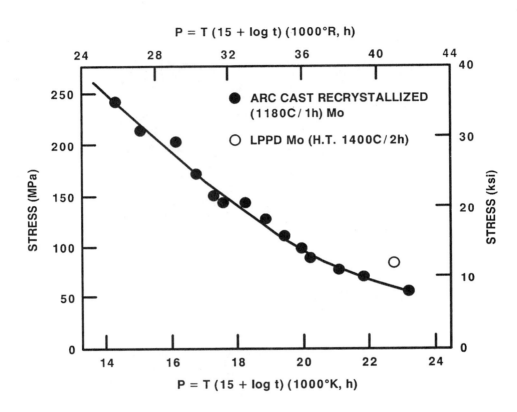

Figure 7   Rupture behavior of elemental Mo deposit.

Near-Net Shape Structures

The tubes from which test bars were made are a demonstration of a very simple shape-making capability. However, a somewhat more complex shape has also been made. A mandrel was machined in a cylindrical shape, varying non-linearly from 4cm diameter at one end, to 7 1/2cm diameter at the other, over a length of 12cm. A deposit was made to a uniform thickness of 0.8 mm using Nb-1Zr powders. The final part is shown in Figure 9 after mandrel removal and heat treatment.

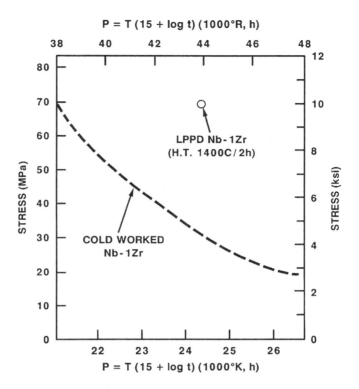

Figure 8   Rupture behavior of Nb-1Zr plasma deposit.

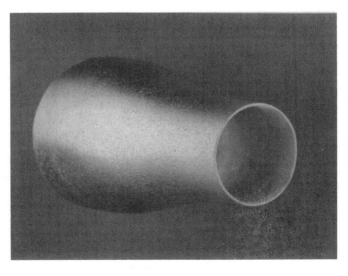

Figure 9   Plasma deposited Nb-1Zr thin-walled contoured shape, 12 cm (4 in.) in length.

Summary

Deposits have been produced in a number of refractory metals and alloys using an RF plasma gun operating in a reduced pressure chamber. Structures of at least 96% of theoretical density have been produced in Mo, Nb, and a number of Nb alloys; attempts using W have not been successful. Under optimum conditions gas contents in the alloy deposits are generally within 150 ppm of the oxygen and nitrogen values measured in the powders.

Thermal treatments, both isothermal and HIP, at 1400°C (2550°F) resulted in some densification, but little change in microstructure. Tensile strengths were comparable to wrought alloy behavior, but ductilities were somewhat lower than expected from wrought material properties. Fractography indicates that hotter plasma conditions and higher temperature thermal treatments are needed to reduce porosity and improve bonding within the structure. Rupture properties appear to be greater than the properties measured for wrought materials.

The near-net shape capability previously demonstrated for RSPD superalloys and composites has been extended to RSPD refractory metals. A thin wall (0.8 mm) contoured tube was produced in Nb-1Zr alloy.

Acknowledgements

The microscopy was performed by J.A. Resue and J. Beale (now at RPI). Gas analysis was measured by S.R. Weissman. Mechanical tests were conducted by C.F. Canestraro, P.L. Dupree and C.P. Palmer. Transmission electron microscopic analyses were performed by N. Lewis and E.L. Hall.

References

1. C. English, "The Physical, Mechanical and Irradiation Behavior of Nb and Nb Base Alloys," *Niobium, Proc. Int. Symp.*, ed. H. Stuart, (Warrendale, PA: TMS, 1984), 239-324.

2. R. Ammon, "Refractory Metals Committee," *J. of Met.*, 39 (8) (1987), 51-52.

3. E.A. Loria, "Nb-Base Superalloys via Powder Metallurgy Technology," *J. of Met.*, 39 (7) (1987), 22-26.

4. D.M. Bowden and P.J. Meschter, "Mechanical Properties and Oxidation Resistance of RS Processed Nb-Based Alloys," (Paper presented at the TMS Fall Meeting, Cincinnati, OH, 1987) 41.

5. C.C. Wojcik, "High Strength P/M Nb Alloys," (Paper presented at the TMS Fall Meeting, Orlando, FL, 1986) 82.

6. C. Austin, General Electric Co., Evendale, OH, private communication with author, 1987.

7. M.R. Jackson, J.R. Rairden, J.S. Smith and R.W. Smith, "Production of Metallurgical Structures by Rapid Solidification Plasma Deposition," *J. of Met.* 33 (11) (1981), 23-27.

8. P.A. Siemers, M.R. Jackson, R.L. Mehan and J.R. Rairden, "Production of Composite Structures by Low-Pressure Plasma Deposition," *Cer. Eng. and Sci. Proc.* 6, (Columbus, OH: Am. Ceram. Soc. 1985), 896-907.

9. M.R. Jackson, P.A. Siemers, J.R. Rairden, R.L. Mehan, and A.M. Ritter, "Composite Structures Produced by Low Pressure Plasma Deposition," *Processing and Properties for Powder Metallurgy Composites*, eds., P. Kumar, K. Vedula and A. Ritter, (Warrendale, PA: TMS, 1988), 45-58.

10. A.M. Ritter and M.R. Jackson, "Microstructural Characterization of RSPD Structures", *Proc. Third Conf. on Rapid Solidification Processing*, (Gaithersburg, MD: NBS, 1982), 270-275.

11. M.R. Jackson and A.M. Ritter, "Chemical and Structural Compatibility in γ/γ-' Composites", (Report 86CRD250, General Electric Co., Schenectady, NY, 1987).

12. J.B. Conway, "Creep-Rupture Data for the Refractory Metals to High Temperatures," (Report GEMP-685 R-69-NSP-9, Cincinnati, OH. 1969).

"RAPID SOLIDIFICATION PROCESSING OF NIOBIUM ALLOYS"

by

S.C. Jha and R. Ray

Marko Materials, Inc.

144 Rangeway Road

N. Billerica, MA 01862, USA

ABSTRACT

Model compositions of niobium alloys were produced by a new rapid solidification process, where rapid solidification was achieved by extracting thin, 20-50 μm diameter filaments from the melt. The rapidly solidified niobium alloy filaments were pulverized, and the powders consolidated by hot isostatic pressing. The grain size of the rapidly solidified alloys after hot-consolidation was in the range of 1-10 μm. Dispersoids of Hf and Nb carbides precipitated uniformly within the grains and at the grain boundaries. The dispersoid particle sizes ranged from 0.3 to 0.01 μm. The rapidly solidified alloys were stronger than corresponding ingot metallurgy commercial WC-103 niobium alloys at low temperatures. However, the high temperature strength of the rapidly solidified alloys fell rapidly, and was comparable to those of WC-103 alloys. Not all pores and voids were eliminated after HIPing, and failure initiated from voids and prior particle boundaries.

Refractory Metals: State-of-the-Art 1988
Edited by P. Kumar and R.L. Ammon
The Minerals, Metals & Materials Society, 1989

INTRODUCTION

Recently, the United States Office of Science and Technology has put forward certain goals for future subsonic, supersonic and transatmospheric flights (1). The thrust is towards designing highly energy efficient systems, which will require the power generating systems to operate at high temperatures. The advanced turbojet and scramjet engines will require materials which can withstand temperatures ranging from 1400K to 2300K. The state-of-the-art superalloys can be utilized to the maximum temperatures of 1350K , and therefore there is an immediate need for viable materials with higher temperature capabilities.

Niobium alloys are known to possess useful strengths at temperatures as high as 1600K. Significant amounts of niobium alloy development activities were undertaken during late 1960's and early 1970's, and several potential alloys compositions were identified (2). However, most of the niobium alloys possessed very poor oxidation resistance at high temperatures. Furthermore, the alloys designed for improved oxidation resistance, proved to be extremely brittle at room temperature and possessed poor workability (3). Due to these adverse results, the interest in niobium alloys declined. However, the recent need for high temperature materials has rejuvenated the interest in niobium alloys, whose workability and formability can be improved by powder metallurgy processing. Niobium is the lightest refractory metal, with density close to that of Ni (8.57 g/cc) and has good thermal conductivity (65.3 W/mK at 873K), therefore they would require lesser air cooling arrangements when utilized in the turbine hot sections.

The success of powder metallurgy niobium alloys will depend on the successful production of high quality niobium alloys powders. The oxidation resistant niobium alloys will contain a large amount of alloying additions and therefore it will be imperative to produce these alloy powders through rapid solidification processing, in order to minimize elemental segregation and formation of coarse intermetallic particles during solidification. Rapid solidification of refractory alloys is difficult, due to the requirements of intense power source and very high temperature to obtain a homogeneous molten metal pool. Recently, Teledyne-Wah-Chang-Albany has built a prototype electron beam atomization process (4), and researchers at McDonnell Douglas Research Laboratories have prepared niobium alloys by electron beam melting and splat quenching (5). Rapid solidification processes based on melt atomization concepts do not attain high enough solidification rates. Rapid quenching by contact with a highly conducting substrate yields the best solidification rates. Marko Materials, Inc. has recently developed a rapid solidification technology for producing rapidly solidified

(RS) filaments of reactive and refractory alloys. The laboratory scale process can produce approximately 500 gms. of RS filaments, and has the potential for development to pilot plant production levels (6).

In this paper the microstructure and mechanical properties of two niobium base alloys produced by Marko's rapid solidification method are described.

EXPERIMENTAL DETAILS

Figure 1 shows the photograph of the rapid solidification equipment for producing RS niobium alloys, installed in Marko's laboratory. Arc-melting is achieved through a non-consumable tungsten electrode. The intended alloy compositions are shown in Table 1. These alloys are a variant of commercial C-103 alloys, and instead of 1 w% Ti, they contain 0.4 and 0.8 w%C additions to generate carbide dispersoids in the matrix. The RS filaments were pulverized to particle sizes below 40 mesh. The fines below 80 mesh were rejected. The flaky powders were introduced in mild steel cans, cold compacted, hot degassed and then hot-upset against a blind die in an extrusion press. The hot upset billets were hot isostatically pressed.

Samples for optical metallography were prepared by standard polishing procedure. A solution of 3 parts HF, 2 parts H_2O_2 and 4 parts H_2O was used as etchant. Samples for transmission electron microscopy were prepared by mechanically grinding niobium alloy samples to 0.1 mm thickness. 3mm discs were punched from the thin sheets. Final electrochemical thinning was performed in a twin jet polishing equipment. The electrolyte used was a solution of 90% CH_3OH and 10% $HClO_4$, cooled to -30°C and polishing was performed at 5.5V. Tensile tests were performed at room temperature, 1144K(1600°F), 1255K(1800°F) and 1477K(2200°F) in air.

Table 1: Chemical Composition of RS Alloys 1 and 2 (nominal) and WC-103.

Alloy	Composition (wt%)
1	Nb-10Hf-0.4C
2	Nb-10Hf-0.8C
WC-103	Nb-10Hf-1Ti-0.7Zr-0.5Ta-0.55W

RESULTS

1 Microstructures

By rapid solidification processing, 20-50 μm thick continuous filaments of alloys 1 and 2 were obtained. The filaments were ductile, and were easily pulverized in a hammer mill. The pulverization took place under the shearing action of hammers in the hammer mill. Very small amounts of fines are sometimes generated by the wear of hammers. These fines were eliminated by rejecting the fraction below 80 mesh.

Figures 2 and 3 are the photomicrographs of alloy 1 in the as HIPed condition. Figure 2 shows the prior particle boundaries, but there are also areas where the prior particle boundaries begin to disappear. In the higher magnification photomicrograph of alloy 1, shown in Figure 3, the grain sizes and morphologies of the rapidly solidified niobium alloys are apparent. The grain sizes are typically in the range of 1 - 10 μm and both equiaxed and columnar grains are evident in the microstructures. A very high density of second phase is observed dispersed both at the grain boundaries and within the matrix. Figure 4 shows the photomicrograph of alloy 2 in the as HIPed state, and show microstructural characteristics similar to those of alloy 1.

Figure 5 shows a typical area of thin foil of alloy 1, viewed in a transmission electron microscope. This micrograph shows the presence of a high density of both coarse and fine dispersoids with particle sizes ranging from 300 nm to 30 nm. These particles are assumed to be carbides of hafnium and niobium. Although no coarsening experiments were performed, it is expected that due to the high diffusivity of C in body centered cubic niobium matrix at elevated temperatures, and abundant supply of carbide forming hafnium and niobium atoms, the coarsening rates will be high. Therefore, further efforts should be made to identify thermodynamically stable and coarsening resistant dispersoids, so that the elevated temperature mechanical properties are maintained during long thermal exposures.

2 Mechanical Properties

Tables 2 and 3 show the tensile test data for alloys 1 and 2 at room and elevated temperatures. These data are compared with the tensile properties of commercially available sheets of WC-103 alloy, which were recrystallized for one hour at 1316K (7). Table 4 shows the tensile test data for WC-103 recrystallized sheets. It should also be noted that WC-103 data is for sheets tested in vacuum, whereas the tensile tests on RS alloys 1 and 2 were carried out in air.

Table 2: Tensile Test Data of RS Alloy 1.

Temperature (K)	0.2% Y.S. (MPa)	U.T.S. (MPa)	% Elong.	% R.A.
300	494.3	587.4	2.8	3.2
1144	270.2	272.3	21.1	35.4

* All tests were carried out in air.

Table 3: Tensile Test Data of RS Alloy 2.

Temperature (K)	0.2% Y.S. (MPa)	U.T.S. (MPa)	% Elong.	% R.A.
300	430.5	692.8	5.8	9.2
1144	289.5	289.5	24.6	37.7
1255	209.6	209.6	25.6	34.7
1477	133.1	192.3	30.1	42.2

* All tests were carried out in air.

Table 4:Tensile Test Data of WC-103 Recrystallized Sheets(7).

Temperature (K)	0.2% Y.S. (MPa)	U.T.S. (MPa)	% Elong.
300	296.0	420.0	27.0
1144	162.0	296.0	18.5
1366	138.0	188.0	45.0
1477	110.0	138.0	-

* Tensile tests carried out in vacuum.

Alloys 1 and 2 are slightly modified compositions of WC-103, in that the WC-103 alloys contain minor additions of Ti, Zr, Ta and W, which have been replaced by 0.4 and 0.8w%C.

An examination of data in tables 2, 3 and 4 reveals that the RS alloys 1 and 2 possess very high room temperature strengths due to their fine grain sizes and high dispersoid content. However, the strength of the rapidly solidified alloys approaches those of the commercial WC-103 alloys at elevated temperatures. Although a direct comparison can not be made, since the WC-103 alloys were tested in vacuum at high temperatures, whereas the rapidly solidified alloys were tested in air. The rapid decline of yield strength of rapidly solidified alloys at elevated temperatures is probably due to their fine grain sizes. However, the high temperature ductilities of the rapidly solidified alloys is comparable to those of WC-103 alloys. The extensive high temperature ductility indicates that there is a significant potential for further alloying to create very high volume fractions of dispersed precipitates in the niobium matrix.

3 Fracture Behavior

The tensile fracture surfaces of room and elevated temperature tensile tested samples were examined by scanning electron microscopy. Figure 6a shows the room temperature tensile fracture surface of RS alloy 1. This fractograph reveals several microcracks and voids in the material, which did not close during hot consolidation and HIPing. These voids and cracks were the primary fracture initiation sites. Figure

6b is a higher magnification fractograph showing that the prior particle boundaries were the weak sites for initiation of the failure, and fracture propagated mostly intergranularly. The ductile yielding of the matrix around the dispersoid particles led to final failure by forming ductile cusps around the incoherent particles, and further coalescence of these ductile cusps.

Figures 7a and b show the tensile fracture surface of RS alloy 2 tested at 1477K (2200°F) in air. The fracture surfaces are heavily oxidized. However, an examination of fractograph in Figure 7a reveals a transgranular failure. The higher magnification fractograph in Figure 7b shows that the final failure occurred through a microvoid coalescence mechanism. The banded nature of the fracture surface is due to the fact that failure initiated at the prior particle boundaries. The deep ductile cusps indicate the presence of a significant amount of plasticity.

DISCUSSION

This experimental study has been one of the first attempts to produce rapidly solidified niobium alloys in large quantities, where melting is achieved through a non-consumable tungsten arc and consequently rapid solidification is attained through melt spinning thin filaments from the molten alloy. Through this process approximately 500 gms. of rapidly solidified niobium alloy filaments were produced per batch. Efforts are currently underway to scale-up this process to 3-5 lbs. batch capacity. The merit of this process is that suitable alloying additions can be made in order to generate a uniform dispersion of various borides, carbides and oxides precipitates. If the sizes of such inert phases can be restricted to 100 nm or smaller, owing to their high strength and niobium at elevated temperatures, they may serve as effective dispersion strengtheners for niobium alloys.

The primary effect of rapid solidification processing is to extend the terminal solid solubility of various solute additions in the solvent matrix. Due to the high cooling rates imparted during rapid solidification, the formation of coarse primary precipitates is inhibited. A fine grained as-cast microstructure is generated by rapid solidification, which is further refined and stabilized by the precipitation of carbide precipitates. Further precipitation occurs, when the matrix supersaturation is relieved during hot consolidation processing of rapidly solidified powders. The key element in the design of dispersion strengthened alloys is to identify solute elements that would lead to the formation of dispersoid which resist coarsening at projected service temperatures. Further experimentation is required to identify coarsening resistant precipitates in niobium base alloys.

The present study has shown that rapidly solidified niobium alloys possess improved microstructural characteristics and can be utilized to develop high performance, oxidation resistant alloys for critical aerospace applications. It has been noted that the niobium alloys designed for better oxidation resistance are extremely brittle and can not be processed by conventional mechanical working techniques(3). Such oxidation resistant alloy compositions can be prepared by rapid solidification techniques to produce fine grain sized alloys which will be virtually free from any large scale segregation. The fine-grained alloys can be processed through powder metallurgical routes to produce near-net shaped products and hardware.

A combination of rapid solidification and powder metallurgy processing of advanced niobium alloys appears to hold significant potential and promise for successful alloy development work. Extensive research will be required to develop the optimum consolidation parameters to produce high integrity bulk alloys and products. Research activities are also required towards understanding the role of rapidly solidified microstructures on the resultant mechanical behavior of the alloys.

SUMMARY

Sufficiently large quantities of rapidly solidified niobium alloys were successfully produced by Marko's rapid solidification process for reactive and refractory alloys. The niobium alloy filaments were pulverized and hot-consolidated. A microstructural and tensile property investigation revealed that the post-hot-consolidated microstructure consists of 1-10 μm grain size, containing a uniform distribution of dispersoids. The carbide precipitates in Nb-Hf-C alloys form by solid state precipitation during hot-processing and are prone to excessive coarsening, due to the high diffusivity of C in Nb matrix. The room temperature tensile strength of RS niobium alloys was in excess of 650 MPa, with adequate ductility. High temperature strength of RS niobium alloys fell rapidly to the levels of conventional ingot metallurgy niobium alloys. An investigation of coarsening resistant and effective dispersion strengtheners in niobium matrix is in order.

ACKNOWLEDGEMENTS

This initial study on developing a rapid solidification process for large scale production of niobium alloys was carried out under the sponsorship of Air Force Systems Command, WPAFB., Ohio. The help and support of Mr. Scott M. Pearl of WPAFB., technical monitor for this program is sincerely appreciated.

REFERENCES

1. M.A. Steinberg, Scientific American, Vol. 255, No. 4, October 1986, pp. 67-72.

2. H. Innoye, "Niobium in High Temperature Applications", Niobium, Proc. of Int. Symposium, ed. H. Stuart, TMS (1984), pp. 615-636.

3. E.A. Loria, Jl. of Metals, Vol. 39, No. 7, July 1987, pp. 22-26.

4. A.F. Condliff, Metal Powder Report, vol. 42, no. 7/8, 1987, pp. 539-541.

5. D.M. Bowden and P.J. Meschter, "Mechanical Properties and Oxidation Resistance of Rapid Solidification Processed Nb-Based Alloys", paper to be presented at 1987 Fall Meeting, TMS-AIME, Cincinnati, OH.

6. Ranjan Ray and Sunil C. Jha, "A New Manufacturing Technology for High Temperature Niobium Base Alloys", Final Report, Contract # F 33657-87-C-2039, Department of the Air Force, August 1987.

7. Robert J. Marsh, Teledyne-Wah-Chang-Albany, private communication.

Figure 1:   Advanced melt spinning equipment installed at Marko's laboratory for rapid solidification processing of refractory alloys.

Figure 2:   Photomicrograph of as HIPed alloy 1.

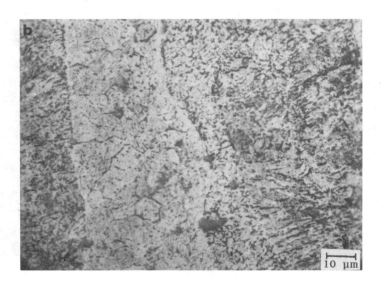

Figure 3:   A higher magnification Photomicrograph of as HIPed alloy 1.

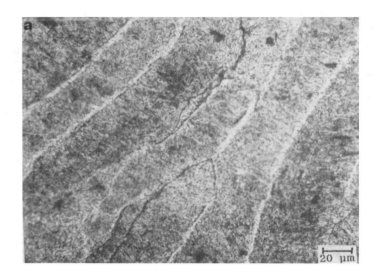

Figure 4:   Photomicrograph of as HIPed alloy 2.

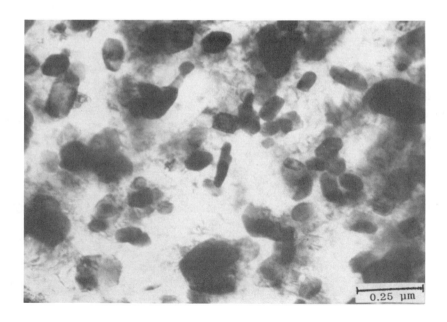

Figure 5:   Transmission electron micrograph of as HIPed alloy 1.

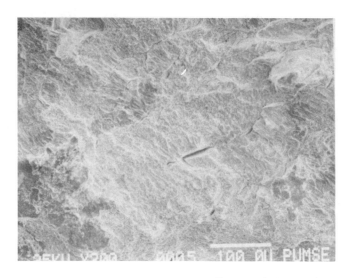

Figure 6a: *Fracture surface of alloy 1, tensile tested at room temperature.*

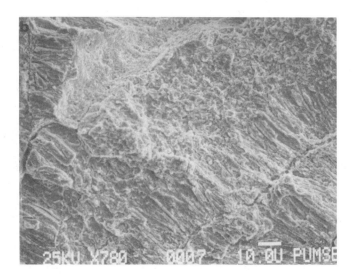

Figure 6b: *Higher magnification fractograph of alloy 1, sample failed in room temperature tensile test.*

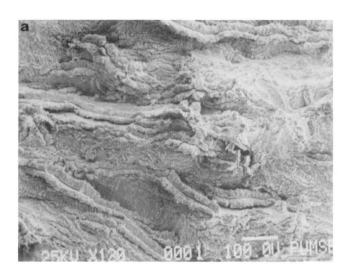

Figure 7a: Fracture surface of alloy 2 tested in tension at 1477 K.

Figure 7b: Higher magnification fractograph of alloy 2 tested in tension at 1677 K.

WELDABILITY AND WELD DOPING OF

COMMERCIAL PURITY MOLYBDENUM

C. V. Robino

Sandia National Laboratories
Albuquerque, NM 87185

Abstract

The GTA weldability and fracture behavior of low and high carbon arc
cast and powder metallurgy molbydenum has been examined. Defect-free welds
were produced in the arc cast grades. For the PM grade, however, severe
centerline cracking and large scale porosity were encountered. The centerline cracking and pore formation were found to result from the dissolution
of oxygen-bearing inclusions present in the starting material. Moreover,
titanium or hafnium additions, incorporated into the weld by RF sputter
coating the joints prior to welding, had several beneficial effects. These
were reductions in the centerline cracking and pore formation in the PM
welds, hardening of the weld fusion zone, and a reduction in the tendency
for intergranular failure in the fusion zone for all base materials. The
mechanisms by which dopant additions improve the welding and fracture
behavior is also discussed.

Refractory Metals: State-of-the-Art 1988
Edited by P. Kumar and R.L. Ammon
The Minerals, Metals & Materials Society, 1989

131

Introduction

The principal problems associated with the use of molybdenum for many applications result from its tendency for brittle intergranular fracture at low temperatures, when in the recrystallized condition. Although not universally accepted as correct, the problem of intergranular failure has been thought to be due to the segregation of impurities, primarily oxygen, to the grain boundaries. Moreover, the tendency for intergranular fracture is particularly severe in molybdenum weldments where other factors can also contribute to poor ductility and joint efficiencies. These factors include contamination during welding, excessive grain growth, and unfavorable weld centerline configurations. Therefore, application of molybdenum is limited by difficulties with both base metal performance and fabrication behavior.

A considerable amount of research on the welding behavior of molybdenum was conducted during the period from 1950-60 (1-5). Most of the work was concerned with developing specific welding procedures and evaluation of the mechanical properties of welds in commercial and developmental materials. A notable exception is the work of Platte (1) who investigated the influence of oxygen on the soundness and ductility of molybdenum gas tungsten arc welds. The results of this work indicated that oxygen in the base metal or in the welding atmosphere can result in weld cracking, porosity, and low ductility. Several strong oxide forming elements were added to experimental heats of molybdenum during vacuum sintering. Of the elements tried, only titanium was found to be a suitable deoxidizing element in sintered molybdenum for welding applications.

More recently, Matsuda et al. (6,7) have examined the weldability and mechanical properties of electron-beam (EB) melted pure molybdenum, arc melted Ti-Zr-Mo (TZM) and powder metallurgy (PM) TZM. It was found that porosity free welds could be made in EB melted molybdenum and arc melted TZM by both electron beam welding (EBW) and gas-tungsten inert gas welding (GTAW). Welds in PM TZM were found to be very porous and this was attributed to the high oxygen content of the PM material.

Other work has addressed the effect of carbon on the welding behavior of molybdenum. Kishore and Kumar (8) evaluated the effect of carbon on EB welds in the PM molybdenum by carburizing samples prior to welding. Carbon additions were found to substantially improve the low temperature ductility. An optimum carbon level of approximately 250 wt. ppm was determined for maximum ductility under the welding conditions used. Kishore and Kumar (8) further argued that the improved ductility imparted by carbon additions resulted from enhancement of the grain boundary crack nucleation stress and that optimum ductilization is achieved when the intergranular crack nucleation stress exceeds the cleavage crack nucleation stress.

Hiraoka et al. (9) conducted a study similar to that of Kishore and Kumar (8), and observed essentially the same results. However, they interpreted the results as indicating that carbon acts to raise the grain boundary crack propagation stress.

Hiraoka and Okada (10,11) have examined the relationships between chemical composition and weld pore formation in PM molybdenum. Their work suggested that weld pore formation results from oxygen containing inclusions within the base metal and suitable deoxidation practice for the powders can result in significant reduction in porosity.

Jellison (12) has examined the dependence of the fracture toughness of molybdenum laser welds on processing parameters and in situ additions of

132

titanium. This work indicated that the fracture toughness of molybdenum welds could be improved by refinement of fusion zone columnar structures and this in turn could be accomplished through the use of pulsed rather than continuous laser power sources. In situ additions of titanium, which were accomplished by vapor depositing titanium on the joints prior to welding, resulted in an increase in the fracture toughness for PM moly and a slight decrease in toughness for both low and high carbon arc cast material. The fracture mode for all grades of molybdenum tended toward transgranular cleavage with increasing titanium additions.

The purpose of the work presented here was to evaluate the effects of impurities, including both tramp elements and intentional additions, on the fabrication and service weldability of molybdenum. This was accomplished through comparisons of the microstructural features and fracture behavior of welds in commercial material both with and without titanium or hafnium additions.

Experimental

Materials

The materials used in this investigation were three grades of commercial purity molybdenum. These three grades are designated by the consolidation procedure used and are: powder metallurgy (PM), low carbon arc cast (LCAC) and high carbon arc cast (HCAC). For each respective grade, all materials used in this study were from the same heat of starting material and were supplied by AMAX Specialty Metals Corporation. Typical impurity analysis and actual carbon and oxygen contents for the experimental materials are given in Table I. Purity for all materials was better than 99.975 wt%. The materials were in the form of 0.051 mm thick sheet and were obtained in the wrought condition.

Welding

The gas-tungsten arc welding (GTAW) method was used for all welds fabricated in this work. Butt welds were made in the 0.5 mm thick sheet by joining two 4.13 cm by 1.91 cm plates along the 4.13 cm edge. The facing edges of the joints were ground flat and perpendicular to the sheet face and all welds were transverse to the rolling direction of the sheet. Prior to welding, the material was cleaned according to a procedure given by Quaglia (13). Following cleaning, the samples were stress relieved at 900°C for 30 minutes in a vacuum of less than 4.7×10^{-6} mbar.

The samples were autogenously welded in an argon filled glove box with continuous circulation and oxygen and water vapor removal systems. These purification systems maintained the oxygen and water vapor levels at less than 1 and 3 volume ppm, respectively. The welding chamber had no provision for nitrogen removal so, prior to the welding of each set of test samples, a minimum of five 5 cm long bead on plate welds on 6 mm thick titanium plate were made. These welds were made in an effort to keep nitrogen levels to a minimum and, since titanium welds discolor at very low impurity levels (approximately 25 ppm total $H_2O + N_2 + O_2$), to insure that the atmosphere purification systems were functioning properly. Molybdenum welds were not made unless all five titanium welds showed no discoloration. Welds were made by a semi-automatic procedure with travel speed held constant and the arc current continuously adjusted during the weld cycle to maintain the bead at 3 mm.

133

Table I. Typical Chemical Analysis of Commercial Purity Molybdenum.
Values given for the PM grade are typical for the starting
powders. Carbon and oxygen values are for the project
materials.

	wt. ppm		
Element	LCAC	HCAC	PM
Aℓ	<25	<16	5-25
Ca	8-23	8-25	3-15
Cd	<8	<15	----
Cu	<5	<5	5-10
Cr	3	4	5-25
Fe	17-25	25	10-100
K	<20	<10	10-30
Mg	<18	<12	1-10
Ni	<5	<5	5-50
Si	<20	<20	5-250
Ta	<15	<15	----
W	---	---	100-300
Zr	---	---	----
All Others	<3	<3	25-75
C	30	220	<10
O	13±4	12	36±8
N	5	3	----
H	3	2	<5

Doped Welds

The effects of titanium and hafnium on the weldability of molybdenum
were evaluated in this work. The elements were deposited on one face and
one edge of the samples after acid cleaning. Both elements were deposited
by RF sputtering and the thickness of the deposits was 5 μm. During
welding the coated edge of the sample was contained in the joint and the
coated face was oriented on the side opposite the arc as shown in Figure 1.
This was done to minimize the effect of the coating on arc efficiency.
Based on coating thickness, the estimated concentration of the dopant
elements in the weld was 0.70 and 2.04 wt% for titanium and hafnium,
respectively.

Analytical

Oxygen, carbon, and nitrogen levels in welded material were determined
by the inert gas fusion technique using a LECO TC136 oxygen/nitrogen
analyzer.

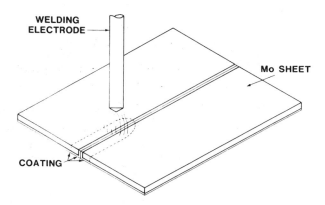

Figure 1 - Schematic diagram showing weld configuration for doped welds.

Fracture surface evaluations were conducted using the scanning electron microscope (SEM). For these observations, either an Etec autoscan or a JEOL model 733 microprobe (EPMA) was used. For most SEM micrographs presented here, the samples were examined uncoated with a low accelerating voltage (5-8 kV). Qualitative maps of elemental distributions on the surfaces of welds were obtained using the SEM mode of the microprobe. For the distribution of light elements, wavelength dispersive x-ray spectrometry was used.

Elemental concentrations on the fracture surfaces of welded samples were obtained with a Physical Electronics Inc. PHI 590 auger electron spectrometer (AES). Samples examined by this technique were fractured within the ultrahigh vacuum chamber at a vacuum of 4×10^{-10} mbar and analyzed as rapidly as possible to minimize beam induced contamination. A 5 kV accelerating voltage and incident beam spot size of approximately 2 μm were used. Estimates of the elemental concentration on the fracture surface were made using a method described by Hondros and Seah (14).

Transmission electron microscopy was used to examine the fine scale structure of the materials processed in this investigation. The foils were produced by electropolishing and were examined in either a Philips EM400 AEM equipped with an EDAX energy dispersive spectrometer and a TN-2000 multichannel analyzer or a Phillips EM430 AEM equipped with an EDAX energy dispersive spectrometer and an EDAX 9900 multichannel analyzer.

Tensile tests were used to evaluate the ductility and fracture characteristics of the welds. All tensile tests on welded material were conducted with smooth bar samples and tests were made in both the longitudinal and transverse orientations relative to the welding direction. The tests were conducted with an Instron screw driven loading frame under constant crosshead speeds. The samples were tested at loading rates of 0.0085 mm/sec. An extensometer with an initial gauge length of 12.7 mm was used to record the sample extension for the entire duration of each test.

For welded material, microhardness traverses across polished sections of the weld were used to characterize the hardness distribution across the weld zone. These tests were conducted using a LECO M-400FT microhardness tester with a diamond pyramid indentor, a load of 200 gm, and a loading time of 15 seconds.

135

General Weldability--Low and High Carbon Arc Cast

In general, the fabrication weldability of both the LCAC and HCAC grades was found to be very good. Figure 2 shows the as-welded surface of a LCAC molybdenum gas-tungsten arc weld. The columnar solidification structure is clearly visible. Note that some of the boundaries in Figure 2 appear separated. Grinding and polishing of the surface revealed, however, that the boundaries were not cracked. The areas of dark contrast between

Figure 2 - Optical micrograph of as-welded top surface of LCAC molybdenum. Fusion line is at bottom. Welding direction is from left to right.

the columns is due to relief between adjacent grains which forms during solidification. In spite of the relatively high restraint for the joint configuration, weld cracking was never observed for either the LCAC or HCAC grades. The surfaces of LCAC welds were very flat, except for slight surface rippling which is just perceptible in Figure 2. The surface of HCAC welds was similar to the LCAC welds except for a slight increase in surface rippling. The column spacing (taken as the average width of the grains at one half the distance from the fusion line to the weld centerline) was essentially the same for both grades at 340-390 μm. Although this spacing is rather large, it is not unexpected for GTA welding of a relatively pure material.

TEM observations for the LCAC and HCAC welds are summarized in Figure 3. Small platelike precipitates are observed in the LCAC grade. Precipitation in the HCAC grade is seen to consist of blocky, occasionally large phases. Convergent beam electron diffraction (CBED) was used to identify these phases as Mo_2C. Precipitation along grain boundaries was observed occasionally in the LCAC grade and extensively in the HCAC grade.

Chemical analysis for carbon indicated 9 and 161 wt. ppm for the LCAC and HCAC welds, respectively. In addition, the measured bulk oxygen contents of the LCAC and HCAC welds were found to be 2 and 6 wt. ppm, respectively. Comparison of these oxygen and carbon levels with those of

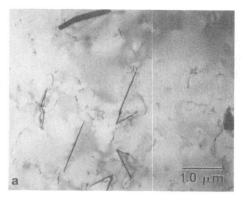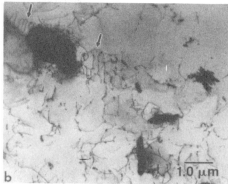

Figure 3 - TEM micrographs showing carbides in molybdenum welds. (a) LCAC, (b) HCAC.

the starting materials, Table I, shows that both deoxidation and decarburization occur during welding. Auger electron spectroscopy from a grain boundary fracture surface of an HCAC weld indicated the presence of carbon, oxygen, and nitrogen. The carbon is not unexpected since, as described previously, TEM evaluation of the HCAC welds showed extensive precipitation along grain boundaries which electron diffraction indicated to be Mo_2C. The oxygen peak was similar in size to that observed (15) to be due to contamination during analysis, although some oxygen is undoubtedly present at the boundaries. The nitrogen peak was somewhat unexpected, since nitrogen was never observed in AES studies of the base materials. It is therefore apparent that some nitrogen contamination occurred during welding. As described previously, the welding chamber had no provision for nitrogen removal so several welds were made on pure titanium prior to the welding of molybdenum samples. Apparently this was insufficient for complete removal of nitrogen from the welding atmosphere. Chemical analysis indicated that the amount of nitrogen contamination was similar for both LCAC and HCAC welds at approximately 20-25 wt. ppm.

In terms of precipitation, the behavior of carbon and nitrogen in molybdenum is similar (16) although the structures of Mo_2C and Mo_2N are different. Mo_2C forms in a hexagonal structure with a c/a ratio of 1.577 while Mo_2N is cubic with a = 4.163 Å (16). All electron diffraction experiments indicated a hexagonal form for precipitates in the arc cast grades. Therefore, it is believed that precipitates in the arc cast grades are molybdenum carbonitrides based on the Mo_2C hexagonal structure.

The dislocation density for both materials was observed to be low although the dislocation density for the HCAC appeared, in general, to be somewhat higher. As shown in Figure 3(b), dislocations are apparently related to the presence of the carbides. This observation supports one of the conclusions of Kumar and Eyre (17,18) that part of the contribution of carbon in increasing the ductility of molybdenum is through generation of dislocations at carbides.

General Weldability - Powder Metallurgy Grade

The fabrication weldability of PM molybdenum was found to be generally poor. Figure 4 shows the as-welded surface of a butt weld in PM

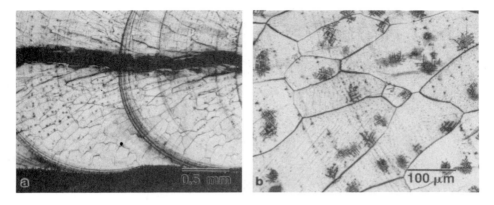

Figure 4 - (a) and (b) Optical micrographs of as-welded top surface of PM molybdenum showing centerline cracking and surface phases.

molybdenum. A severe through thickness crack, which ran the entire length of the sample, is observed at the weld centerline. In comparison with the LCAC weld, Figure 2, several other features are noted. First, a dendritic phase is seen to be precipitated on most of the weld surface. The surface of the weld is also more rippled than the LCAC grade and occasionally there are very large ripples observed. Finally, the grain size is much smaller and the columns do not extend from the fusion line to the centerline, as is typical for the LCAC.

During welding of the PM grade it was observed that approximately 10-15% higher welding currents were required to maintain the same bead width as the LCAC welds. Also, the arc was observed to be very erratic and occasionally (roughly 4-5 times per sample) a large gas bubble was evolved from the weld puddle just ahead of the solidification front. A section showing the size of the resulting pore is presented in Figure 5.

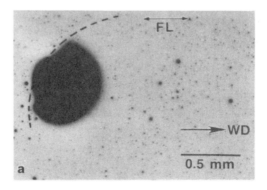

Figure 5 - Optical micrograph of PM weld sectioned in the plane of the sheet. The dotted line to the left of the pore indicates the approximate location of the solidification front when the pore formed. FL and WD indicate fusion line and welding direction, respectively. Also shown is the general pore distribution in PM welds.

The surface phases of Figure 4(b) were evaluated using WDS mapping techniques as shown in Figure 6. The dendritic features are seen to contain calcium, oxygen, and to a lesser extent, aluminum. Aluminum is not observed to be present in the nodular surface phases indicated by arrows in Figure 6(a). Both the nodular and dendritic phases were observed to be quite thin as low accelerating voltages were required to observe them in the backscattered electron imaging mode. Figure 6 also shows that, at the surface of the weld, the grain boundaries are decorated with similar phases. The grain boundary is indicated by the arrow in Figure 6(a).

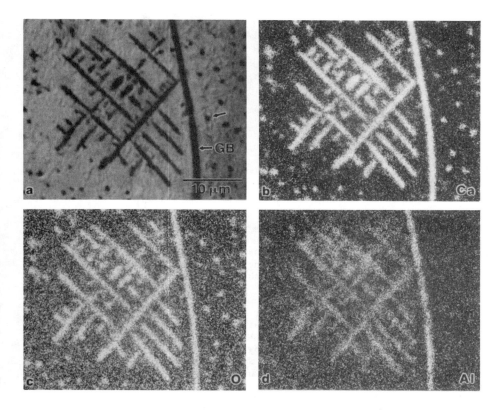

Figure 6 - (a) Backscattered electron image of surface phases on PM welds. Note dendritic phase and nodular phase indicated by arrows. (b-d) WDS x-ray maps of area in (a) for Ca, O, and Aℓ, respectively.

From these observations, a mechanism for the formation of weld porosity and the cracking tendency for PM molybdenum can be described. It has been shown previously (15) that, because of the very low solid solubility of oxygen in molybdenum, the bulk oxygen content of PM molybdenum is controlled by the inclusions present in the starting material, and an example of such inclusions is shown in Figure 7. It therefore seems apparent that the poor weldability of this material can also be attributed to these same inclusions. It must be noted that the melting points for Ca-Aℓ-O compounds fall within the range of 1500-1600°C compared to 2610°C for molybdenum.

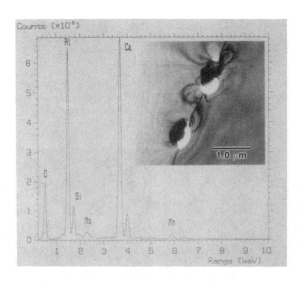

Figure 7 - TEM micrograph and EDS spectra of inclusions in PM molybdenum showing that the inclusions are oxygen bearing.

Therefore, it is expected that during welding, the inclusions dissolve and a significant quantity of oxygen is taken into solution in the liquid molybdenum. In fact, it is this process that is used to remove oxygen and other impurities in the vacuum arc cast grades. In the following discussion, the dissolution of oxide bearing inclusions is used to describe the observations of poor fabrication weldability of PM molybdenum.

Effect on Heat Input. The welding current required for the PM grade was significantly higher than that for the arc cast grade and it is thought that the reason for this is the liberation of impurities into the arc. The liberation of oxygen from the weld puddle will locally change the composition of the welding atmosphere thereby altering the arc efficiency. This does not, however, fully account for the observation of poor arc stability since oxygen is commonly added to argon for gas-metal arc welding to improve arc stability and improve arc efficiency (19). The effect of calcium and/or aluminum vapor or oxides of these elements in the vapor phase is not known, but it is likely that intermittent liberation of these species into the arc would affect the arc stability.

Surface Phases. The surface phases are of similar composition to the initial inclusions present in the material (see Figure 7) so it is clear that their origin is from the bulk inclusions. Their presence on the surface is important in that it indicates that some purification of the molybdenum occurs during welding, i.e., some calcium and aluminum are removed from the bulk. However, the degree of this removal is thought to be small. The precipitates are apparently not grown by condensation from the vapor phase as they are only observed to occur on the fusion zone of the weld and not on the adjacent molybdenum or the weld fixtures. They also do not appear to be related crystallographically to the molybdenum since many are observed to cross grain boundaries as seen in Figure 4(b) without change in their morphology. The mechanism by which the surface phases assume the dendritic morphology is not fully understood.

140

<u>Pore Formation and Weld Cracking</u>. Hiraoka and Okada (11) have recently
related pore formation in electron beam PM welds to the presence of oxide
forming elements. Through variations in powder processing they were able
to indirectly show that calcium and magnesium oxides were the primary
species responsible for pore formation. Magnesium was not observed to a
significant extent in the present work but calcium and aluminum oxides have
been shown to be related to pore formation, since molybdenum with low
inclusion contents (LCAC) does not show weld porosity. The importance of
this observation is subtle but significant. It has been known for some
time (see for example Refs. 1 and 2) that high oxygen contents in molybdenum contribute to weld pore formation, but the underlying reasons for the
high oxygen content of PM molybdenum have previously been somewhat vague.
That is, most authors have simply argued that the oxygen content of PM
molybdenum is high and therefore it is more prone toward porosity. The
results of this investigation and the work by Hiraoka and Okada (11)
suggest that it is the secondary impurities, such as calcium, which control
the bulk oxygen content of the PM molybdenum. Therefore, efforts to
improve the weldability of PM molybdenum should not be directed at removal
of oxygen per se, but rather at removal of secondary impurities. Hiraoka
and Okada (11) show ways in which this can be accomplished through
refinement of the powder production process. Subsequent sections of this
paper will illustrate methods by which this can be accomplished during the
welding process.

The tendency for weld cracking in the PM grade material is also related
to the presence of secondary impurities. Figure 8 shows the surface of a

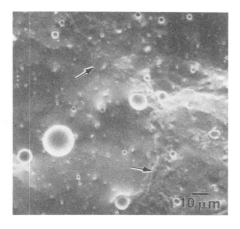

Figure 8 - SEM micrograph of centerline crack surface of PM molybdenum
weld.

weld centerline crack in PM molybdenum. Cracking is seen to be
intergranular and there is no evidence for liquid phases having been
present when cracking occurred. There is a great deal of porosity and
inclusions are also apparent at the boundary surface. Clearly the high
concentration of both pores and inclusions in the boundary increase the
tendency for cracking as stress develops during weld cooling. Auger
analysis of grain boundary fracture surfaces of PM welds indicated the

presence of calcium (as expected from the inclusion content) as well as a significant amount of oxygen. Nitrogen was also detected as described previously for the LCAC welds. Figure 6 shows that, at the surface of the weld, the grain boundaries are decorated with a continuous layer of oxide. Thus, it is apparent that the presence of calcium, and to a lesser extent aluminum, can have both direct and indirect consequences with respect to the cracking tendency of PM welds. The indirect consequence of their presence is the liberation of oxygen during welding which results in excessive porosity as well as embrittlement. The direct consequence of their presence is the possible formation of boundary films which further increase the tendency for cracking. Evidence for such films can be seen in Figure 8.

Tensile and Microhardness Tests

Tensile tests were conducted on welds in LCAC and HCAC grade molybdenum and the results of these tests are given in Table II. As shown, tests were conducted in both longitudinal (weld direction parallel to tensile axis) and transverse (weld perpendicular to tensile axis) orientations. In some cases for the transverse tests two values for the percent elongation are given. The first value corresponds to the elongation of the weldment as a whole and is measured over a full 12.7 mm gage length. The values given in parentheses are the elongation of the fusion zone alone and were obtained by measuring the full width of the fusion zone before and after testing. Tests were not conducted on the PM grade welds because of the centerline cracking problem described previously.

Table II. Summary of Tensile Test Results for Undoped Molybdenum Welds.

Material and Orientation	σ_Y[1] MPa	σ_{TS}[2] MPa	%EL[3]
LCAC (L)	387	433	6.0
LCAC (L)	---	419	0.0
LCAC (T)	400	420	3.2
LCAC (T)	419	434	2.2
LCAC (T)	---	443	1.0(2.6)
HCAC (L)	424	474	8.3
HCAC (T)	428	480	4.8(13.7)
HCAC (T)	454	502	4.9(12.5)

(1) For samples which indicated a yield point, upper yield point given.
(2) Fracture location for all transverse tests was at or near weld centerline.
(3) Values given in parentheses are for % elongation measured across fusion zone only.

Relative to values obtained for recrystallized molybdenum, the strength levels for the welds are quite high. The tensile strength values for the LCAC welds are roughly 90% of that observed for recrystallized molybdenum (15) while yield strengths for the LCAC welds are generally higher than that for purified material, probably as a result of the higher interstitial content of the welds. The high carbon material shows higher strength as well as higher ductility than the low carbon grade. These results are in accord with other investigations (8,9) of the effect of carbon on the ductility of molybdenum welds. The fusion zone ductility for HCAC transverse tests exceeds that for the overall gage length by more than a factor of two, indicating that significant strain localization (20) occurs in the transverse tests. This localization occurs as a result of the large grain size difference, and thus strength difference, between the weld zone and the base metal. The point is further illustrated by the microhardness traverse for an LCAC weld which is presented in Figure 9. Although it was not possible to conduct tensile tests on the PM grade, microhardness traverses showed that hardness variations for the PM material are similar to those for the arc cast grades.

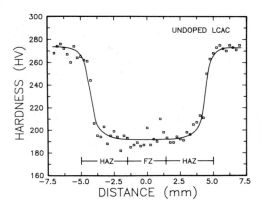

Figure 9 - Microhardness traverse for undoped LCAC weld.

Fractography of Undoped Welds

The results of SEM evaluation of the tensile fractures of welds in the arc cast grades are presented in Figure 10. Fracture in the LCAC grade transverse orientation was essentially all intergranular and occurred along the centerline of the weld. The fracture paths in longitudinally oriented LCAC welds were generally flat cleavage planes perpendicular to the tensile axis, although some intergranular steps between the cleavage planes in individual grains are observed, Figures 10(c) and (d). This is expected since, as shown in Figure 2, grain boundaries at the weld centerline of longitudinal tests are oriented nearly parallel to the loading axis. Fracture in the HAZ of longitudinally oriented samples was primarily intergranular.

The results for the HCAC welds were similar to those observed for the LCAC, except that significant amounts of cleavage are observed for the transverse tests. Fracture for the transverse HCAC was also near the weld centerline. The higher percentage of cleavage fracture in the HCAC grade

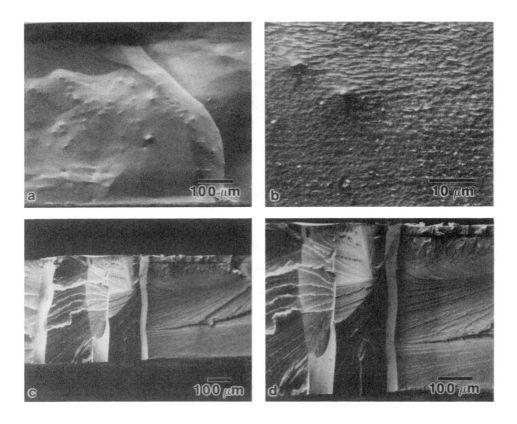

Figure 10 - SEM micrographs of tensile fractures of LCAC welds.   (a) and
(b) traverse orientation, (c) and (d) longitudinal orientation.

correlates well with the higher ductility of the HCAC welds (especially if
the fusion zone elongation is compared, Table II).   It is therefore
apparent that the presence of carbon reduces the tendency for intergranular
fracture at room temperature, as has been observed by others (8,9).
Comparison of the intergranular fracture surfaces for the low and high
carbon grades showed that although grain boundary precipitation was evident
for both grades, it was much more extensive in the high carbon material.

Effect of Carbon

Kumar and Eyre (17) have proposed that the effect of carbon on
improving the ductility of molybdenum has three main parts.   First, they
found that the presence of carbon reduced the free energy of segregation of
oxygen, thereby lowering oxygen segregation levels.   Second, carbon was
found to form misfitting Mo_2C precipitates which can act as dislocation
sources.   Third, the carbides which were present on grain boundaries were
observed to form semi-coherent interfaces with one or both grains, thereby
reducing the grain boundary energy.   However, they (17) concluded that the
first of these effects was the major factor influencing ductility.

144

All three of these mechanisms seem plausible, and were observed in varying degrees during the course of this work. For molybdenum welds in particular, a fourth possible effect was noted previously. That is, carbon improves the ductility of molybdenum welds by deoxidation (or prevention of oxygen contamination) during welding. It is difficult to determine, for polycrystalline material, which of these mechanisms predominates. However, the dominant feature on the grain boundaries of high carbon samples is precipitation. Thus, it is clear that for high carbon welds the boundary structure is significantly different than low carbon welds or pure recrystallized material. Although this structural difference is chemical in nature (i.e., carbide precipitation), it appears that structural features, such as dislocation generation at the carbides, must predominate. This is also in agreement with previous work (15), which indicated that reductions in uncombined oxygen concentrations to below detectable levels was insufficient to inhibit grain boundary fracture.

Behavior of Doped Welds

The primary characteristic dictating the choice of dopants was the oxide forming characteristics of the prospective dopants. For example, titanium and hafnium both have significantly more negative free energies of formation for their respective oxides than that for molybdenum oxide. Titanium additions had also been previously shown (12) to improve the fracture characteristics of molybdenum welds. Further, these elements are relatively easy to deposit by conventional sputtering techniques, so that their introduction into the weld zone is straightforward and can be adapted to realistic engineering applications.

General Weldability - Titanium and Hafnium Additions

The welding of titanium and hafnium doped materials are discussed in this section. These elements were found to result in similar improvements in weldability and mechanical behavior, although, in general, the effects of hafnium were somewhat more pronounced. Therefore, emphasis in this section is placed upon the hafnium additions, except where significant differences were observed between the effects of the two elements. In the following sections, abbreviations are used to describe the various dopant/base metal combinations; with TI or HF indicating titanium or hafnium additions and PM, LC, or HC representing powder metallurgy, low carbon arc cast and high carbon arc cast base metals.

Fusion Zone Structures. Optical micrographs showing the effects of titanium and hafnium on the bulk structure of LCAC and PM welds are presented in Figures 11 and 12. Inspection of this figure indicates that the effect of titanium on the grain structure of the arc cast grades is minimal. Titanium additions have altered the etching behavior of the welds as the fusion line is clearly visible in the TILC weld. For the TIPM weld, titanium is seen to reduce the tendency for centerline cracking but has not eliminated the formation of large pores, although general porosity is reduced compared to the undoped material. In contrast, hafnium additions are seen to significantly refine the grain structure of the arc cast grade. Hafnium additions are also seen to eliminate weld centerline cracking and large porosity in the PM material.

TEM evaluation of the doped welds indicated that significant precipitation had occurred during the weld cycle, especially in the hafnium doped material. In general, precipitation in the doped welds was seen to be very

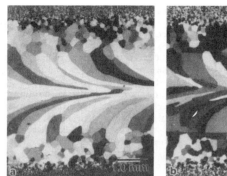

Figure 11 - Optical micrographs comparing grain structure of LCAC welds. (a) undoped, (b) titanium doped, (c) hafnium doped.

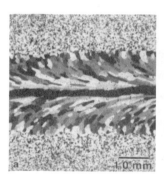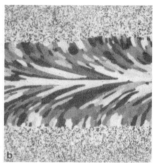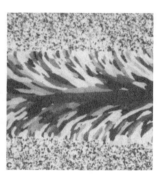

Figure 12 - Optical micrographs comparing grain structure of PM welds. (a) undoped, (b) titanium doped, (c) hafnium doped.

inhomogeneous, and various stages of the precipitation process were often observed within the same weld and occasionally within the same foil. An example of this is illustrated in the micrographs of Figure 13. Figure 13(a) shows a cluster of plate shaped precipitates adjacent to a relatively precipitate free area in a HFLC weld. The distribution of precipitates in Figure 13(b) suggests that cellular solidification occurs in the hafnium doped welds and was observed in both HFLC and HFHC samples. Cellular solidification has been observed by Wadsworth (20) in GTA welds in TZM. Titanium doped welds did not exhibit cellular solidification patterns and this is thought to be related to the observation that a smaller amount of titanium (relative to hafnium) is incorporated into the weld zone. This observation is discussed more fully below. Attempts to identify the precipitates through selected area electron diffraction were not successful as the observed lattice parameters showed poor matching with any of the commonly observed (16) carbides or nitrides in ternary Mo-Hf-(C,N) systems. Problems with precipitate identification were also encountered in the

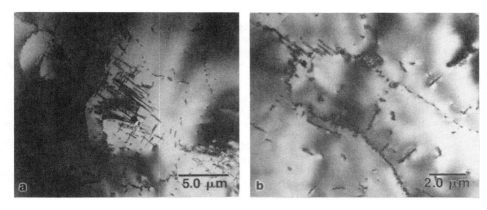

Figure 13 - Bright field TEM micrographs of HFLC welds showing (a) clustering of precipitate plates, (b) cellular solidification pattern.

titanium doped welds where similar precipitation was observed. Precipitation in molybdenum containing carbon and hafnium or titanium has previously been observed to be complex (16,21,22). Ryan (16) has shown that for isothermal aging, overlapping temperature stability ranges exist for the various possible carbides (i.e., Mo_2C, TiC, HfC) in these systems. X-ray fluorescence measurements on extracted precipitates (22) indicated that molybdenum can substitute for titanium in TiC and the degree of substitution is dependent on alloy composition as well as processing conditions. For the welds considered here, the complexity of the precipitation reactions is further increased by nitrogen contamination, nonequilibrium solidification, and the relatively high cooling rates experienced during welding.

The grain boundaries of the doped welds displayed significant amounts of precipitation, as shown in Figure 14. In general the precipitates at the boundary show low energy interfaces with at least one of the grains. As noted previously, this has been suggested (8) as a cause for the lower tendency for intergranular fracture in carbon containing molybdenum. This possibility is discussed in more detail in subsequent sections.

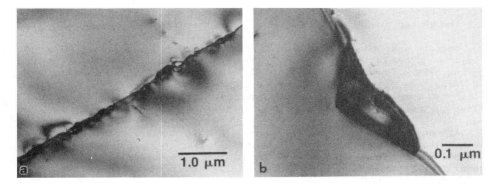

Figure 14 - TEM micrographs of grain boundary precipitates in HFHC weld showing apparent partial coherency of the precipitate with one grain.

147

Weld Zone Deoxidation.   Titanium and hafnium additions were found to promote deoxidation of the welds.   It is important to note that in the context of welding, deoxidation as discussed in this paper pertains to both removal of uncombined oxygen from the base materials and prevention of oxygen contamination from the welding environment.

The mechanism of deoxidation was found to be composed essentially of two parts which occur to greater or lesser degrees depending on baseplate composition and dopant used.   The first of these is the formation of oxygen bearing phases on the surface of the welds and is similar to that observed for undoped PM welds (see previous sections).   The second is the formation of stable oxygen bearing inclusions within the weld metal itself.   Inclusions containing the dopant additions were observed in all baseplate materials.   Examples of inclusions in HFLC welds are visible in Figure 13. Inclusions in the titanium doped welds were generally spherical in appearance while those in the hafnium doped were occasionally faceted.   EDS analysis of inclusions in the arc cast grades invariably showed only the presence of the dopant element although it is possible that molybdenum is included within the oxide.   As might be expected from the initial oxygen contents of the baseplates, Table I, inclusions in the PM welds were more numerous than in the arc cast welds and were occasionally very large.   EDS analysis often indicated calcium, as well as the dopant additions, in the PM weld inclusions.

The formation of surface phases was strongly dependent on the baseplate composition and dopant used.   In addition, the morphology and composition of the surface phases is also a function of the baseplate material. Comparing first the effect of titanium for the TIPM weld, the WDS analysis showed that both titanium and calcium were contained in the surface phases. Thus, in the PM welds, titanium appears to play a dual role as both a deoxidant and as a flux for removal of calcium.   As was noted earlier, calcium oxides are formed on the surfaces of PM welds even without titanium additions.   However, surface phase formation in the TIPM welds was found to be more extensive than in the UDPM welds.   Further, although no comprehensive comparison was made, AES analysis of grain boundary fracture surfaces in UDPM and TIPM welds generally showed lower calcium contents for the TIPM welds.

Titanium additions to LCAC welds result in surface phases which are essentially titanium oxide.   Bulk chemical analysis for the oxygen content of the doped welds indicated an oxygen content of 16 wt. ppm for TILC and 23 wt. ppm for the HFLC weld.   These values are higher than those observed for the undoped welds but are similar to that of the starting baseplate and are thought to be due to the formation of inclusions in the weld.

Surface phases were not commonly observed in hafnium doped welds except for the HFPM welds.   Relative to the titanium doped welds, surface phases on the HFPM welds were quite rare.   This is thought to be a consequence of the densities of the specific oxides.   Hafnia has a density of 9.68 g/cm^3 which is close to that of molybdenum and thus would not be expected to float to the surface of the weld pool as readily as titania (4.26 g/cm^3).

AES analysis of the grain boundaries in HFLC and TILC welds invariably showed the presence of oxygen in amounts similar to that observed for the undoped welds.   Thus, although the dopants are seen to reduce the available quantity of uncombined oxygen (especially in the doped PM welds where porosity is reduced), any possible changes in the oxygen content at the grain boundaries could not be detected.

148

Role of Dopants in Fabrication Weldability. An important observation of the preceding sections was the effect of titanium and hafnium on eliminating the tendency for weld centerline cracking in PM molybdenum welds. Earlier it was shown that weld centerline cracking and large pore formation were a consequence of the dissolution of inclusions present in the PM baseplate. As shown previously (15), essentially all of the oxygen in the PM grade prior to welding is contained in inclusions. Therefore, there is a stoichiometric amount of reactive solute (i.e., Ca, Aℓ) present to react with oxygen during welding. Thus, it is unlikely that during the relatively short time associated with the welding process, complete reaction can occur to reform the inclusions. In the doped case, however, there is a significant excess of strong oxide forming elements present in the weld pool. Thus, more nearly complete combination of the oxygen can occur.

On a molar basis, the amounts of titanium and hafnium deposited on the weld samples are similar. However, it is thought that the amounts of each respective dopant which are actually incorporated into the weld are significantly different. This conclusion was reached on the basis of the following observations. First, the quantity of material vaporized during the welding process (as indicated by the amount of material condensed on the weld fixtures after welding) was much higher for titanium doped samples than for hafnium doped. In fact, essentially no vaporization was observed for the hafnium case. Second, since the melting point of hafnium (2230°C) is closer to that of molybdenum than titanium (m.p. = 1812°C) it is expected that there is less melting of the hafnium coating ahead of the molten pool. The importance of this is that if the coating does not melt significantly far ahead of the weld pool, there is less chance that the molten dopant can spread away from the weld zone. The formation of less large scale weld porosity in the HFPM weld as compared with the TIPM, Figure 12, also indicates that more hafnium is taken into the weld than titanium.

Differences in the amounts of hafnium and titanium contained in the welds are also apparent in the solidification structure of the welds. For hafnium additions, cellular solidification was observed while this was not the case for titanium doped welds. This is thought to be due primarily to differences in the amount of dopant taken into the weld, and not differences in solubility, since GTA welds in TZM (0.5 wt.% Ti) typically show cellular solidification (20).

The refinement of the weld zone grain size by hafnium additions, Figure 11(c), is an important result since it shows that significant reductions in the normally very large grain size of molybdenum welds can be achieved through simple chemical means. Further, this refinement tends to reduce the occurrence of unfavorable weld centerline configurations observed in undoped welds. The mechanism of refinement is not entirely clear, but is thought to be related to rejection of solute ahead of the solidification front (i.e., hindrance of growth) rather than an inoculation effect. Although no data is available on the solubility of hafnium oxide (m.p. = 2760°C) in liquid molybdenum, if it is considered to be essentially inert than it might be expected to form at temperatures above the melting point of molybdenum, and therefore be a possible nucleation site for new grains. However, as shown in Figures 13(a) and (b), the hafnium containing inclusions are invariably observed at solidification cell edges rather than their centers or at grain boundaries as might be expected if they had acted as nucleation sites for new grains.

Mechanical Behavior

Tensile tests of the doped welds proved to be somewhat inconclusive as longitudinal tests invariably fractured at stresses below that observed for undoped welds. However, in these cases fracture was always observed to have been initiated at a fusion line cavity. These cavities were formed occasionally in the baseplate along the fusion line and were apparently caused by melting of the dopant coating adjacent to the weld. These regions are thought to alloy with the baseplate thereby forming a comparatively low melting temperature liquid pocket, which is subsequently pulled into the fusion zone.

Failure for the transverse tests on both titanium and hafnium doped samples always occurred in the HAZ. For these tests, fracture did not always initiate at fusion line cavities since it was possible to machine the samples from regions of the weld where fusion line cavities were minimal. Thus, doping of the welds is seen to have shifted the fracture sensitive region of the weldment from the fusion zone to the HAZ.

Microhardness profiles for the hafnium doped LCAC ‚baseplate are presented in Figure 15 and are representative of all dopant/baseplate combinations. Comparison of these curves with those obtained for the

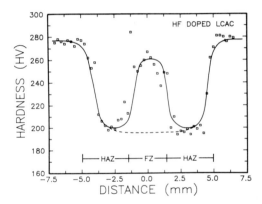

Figure 15 - Microhardness traverse for hafnium doped LCAC weld.

undoped welds shows that both titanium and hafnium result in significant hardening of the weld fusion zone. The hardness of the doped welds approximates that of the stress relieved baseplate and were generally higher for Hf than Ti doped. The increases in fusion zone hardness are clearly due to precipitation as described earlier, although some increment of hardening in the hafnium doped arc cast grades is due to grain refinement. Considering the hardness data and the effects of strain localization, it is expected that fracture in the transverse tests should occur in the HAZ, as was observed.

Fractography

SEM observations of the fracture surfaces of the doped LCAC weld are shown in Figure 16 and represent the behavior for all baseplates. Micrographs are given for both dopant types and longitudinal and transverse

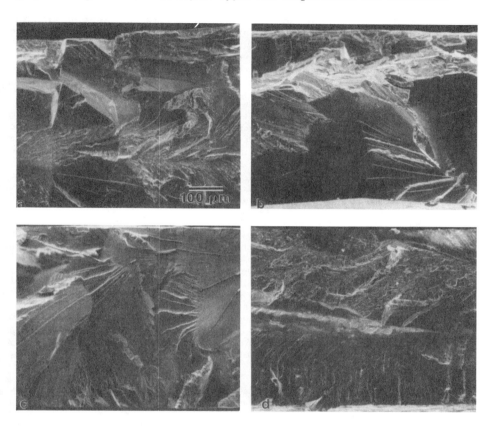

Figure 16 - SEM micrograph of fracture surfaces of doped LCAC molybdenum welds. (a) longitudinal TILC, (b) transverse (cantilever bending) TILC, (c) longitudinal HFLC, (d) transverse (cantilever bending) HFLC.

orientations. Since the transverse tensile samples for the doped welds did not fracture in the fusion zone, the fractures shown in Figure 16 were produced by clamping the samples in a vice along the centerline and using a hammer to break the samples by cantilever bending. In this way the fracture was forced to occur near the weld centerline. Essentially no plastic deformation was observed for these tests.

The longitudinal test for titanium additions in LCAC is seen to be similar to its undoped counterpart, Figure 10(d), in that the fracture mode is generally transgranular with occasional grain boundary fracture steps between grains. Fracture of the TIPM sample was also similar. For the TIHC weld, no intergranular fracture is observed between the cleavage facets of adjacent grains.

Transverse (cantilever bending) tests in all titanium doped samples show significant amounts of cleavage fracture, approximately 60-70%, in spite of the relatively high severity of the test. For hafnium doped samples in both test orientations, the fracture mode was observed to be essentially all transgranular cleavage.

The results of this part of the investigation are in agreement with the work of Jellison (12) in that titanium additions are seen to reduce the tendency for intergranular fracture. In addition, hafnium is seen to act in the same way and is more effective, under the conditions of the study, in promoting cleavage fracture.

Role of Dopants

The role of these dopants in altering the fracture mode of molybdenum is believed to consist of several parts. First, the dopant additions are seen to reduce the available quantity of uncombined oxygen (both from the base metal and welding atmosphere) as evidenced by the inclusions in the weld zone and the improvements in PM weldability. Second, the dopants combine with carbon from the base metal (and nitrogen from the welding atmosphere) resulting in precipitation hardening of the fusion zone. This hardening, as well as grain refinement for the hafnium doped samples, reduces the effects of strain localization in the fusion zone and thus shifts the fracture prone region to the HAZ. Finally, the dopant additions are seen to precipitate extensively along grain boundaries in the fusion zone, which, due to the nature of the precipitates, provides an increment of strengthening and reduces grain boundary fracture by the mechanism of Kumar and Eyre (18). Of these effects, the most important is apparently precipitation along grain boundaries since changes in the oxygen content of the boundary could not be detected and, as shown in previous work (15), reductions in oxygen content at the grain boundaries alone are insufficient to eliminate the tendency for grain boundary fracture. Further, hafnium is seen as a more effective dopant than titanium in this respect and this is a consequence of the fact that hafnium is inherently a more refractory material and thus more is incorporated into the weld using the alloying technique of this study.

In a practical sense, these results are important in that they show that potentially useful joints can be produced in PM welds by simple alloy additions during welding. Although tensile tests could not be used to fully evaluate their behavior, the microhardness and fracture data indicate that improved tensile behavior can be expected for doped welds. The problem of fusion line cavity formation is not viewed as a serious drawback as it should be possible to eliminate the problem by suitable masking of the unmelted areas of the joints during coating. It also seems likely that further improvements in properties should be obtainable by optimization of dopant additions and welding parameters.

Conclusions

The weldability and fracture of molybdenum welds were evaluated and the effects of weld dopants were examined. The major conclusions which can be drawn from this study are:

(1) The fabrication weldability of powder metallurgy molybdenum was observed to be poor. Dissolution of oxygen-bearing inclusions was found to be the underlying factor controlling porosity and centerline cracking in

powder metallurgy welds. Significant reductions in porosity and elimination of centerline cracking were achieved through weld doping with strong oxide forming elements.

(2) Dopant additions of titanium and hafnium were found to increase the hardness of the fusion zone and to shift the fracture location from the fusion zone to the heat-affected zone. These dopant additions also reduced the tendency for intergranular fracture in the weld fusion zone. Although these dopants were found to contribute to fusion zone deoxidation through compound formation, reductions in the tendency for intergranular failure could not be correlated with changes in the oxygen content of the boundaries. The role of dopants in reducing the tendency for intergranular fracture in molybdenum welds was found to result primarily from precipitation of the dopant species as compounds with carbon and nitrogen within the welds, and especially at grain boundaries.

Acknowledgements

The author would like to thank Dr. James L. Jellison for his support during the course of this work. He would also like to thank Dr. John Stephens and Dr. Mike Cieslak for reviewing the manuscript. Thanks are also due to Mr. Jim Kerner and Mr. Dave Calvert for help with the microscopy. This work was performed at Sandia National Laboratories supported by the U.S. Department of Energy under Contract Number DE-AC04-76DP00789, and at Lehigh University.

References

1. W. N. Platte, "Influence of Oxygen on Soundness and Ductility of Molybdenum Welds," Welding Journal, 35 (1956) 369s-381s.

2. W. N. Platte, "Effects of Nitrogen on the Soundness and Ductility of Welds in Molybdenum," Welding Journal, 36 (1957) 301s-306s.

3. J. H. Johnston, H. Udin, and J. Wulff, "Joining of Molybdenum," Welding Journal, 33 (1954) 449s-458s.

4. R. E. Monroe, N. E. Weare, and D. C. Martin, "Fabrication and Welding of Arc-Cast Molybdenum," Welding Journal, 35 (1956) 488s-498s.

5. N. E. Weare, R. E. Monroe, and D. C. Martin, "Ductility of Tungsten-Arc Welds in Molybdenum," Welding Journal, 36 (1957) 291s-444s.

6. F. Matsuda, M. Ushio, and K. Nakata, "Weldability of Molybdenum and Its Alloy Sheet (Report I)," Trans. JWRI, 8(2) (1979), 217-229.

7. F. Matsuda, M. Ushio, and K. Nakata, "Weldability of Molybdenum and Its Alloy Sheet (Report II) - Ductility of EB and GTA Welds," Trans. JWRI, 9(1) (1980) 69-78.

8. R. Kishore and A. Kumar, "Effect of Carbon on the Ductilization of Electron-Beam Welds in Molybdenum," J. Nucl. Mater., 101 (1981) 16-27.

9. Y. Hiraoka, M. Okada, and R. Watanabe, "Low-Temperature Ductility of an Electron-Beam-Welded Molybdenum Sheet Predoped with a Small Amount of Carbon," J. Nucl. Mater., 83 (1979) 305-312.

10. Y. Hiraoka and M. Okada, "Weldability of Powder-Metallurgy Molybdenum with Low Oxygen Content," *Z. Metallkde.,* 78 (1987) 197-200.

11. Y. Hiraoka, H. Irie, M. Okada, and R. Watanabe, "Bend Properties of Electron Beam-Welded Sintered Molybdenum Sheet," *Powder Met. Int.,* 14(3) (1982) 148-152.

12. J. L. Jellison, "Dependence of Fracture Toughness of Molybdenum Laser Welds on Dendritic Spacing and In Situ Titanium Additions," (Report SAND79-0193, Sandia National Laboratories, 1979).

13. L. Quaglia, "Surface Treatment of Non-Ferrous Metals for the Purpose of Gas Analysis," (BCR Information, Commission of the European Communities, 1979).

14. E. D. Hondros and M. P. Seah, "Segregation to Interfaces," *Int. Met. Rev.,* 222 (1977) 262-301.

15. C. V. Robino, "Chemical Effects on the Weldability and Fracture of Molybdenum," (Ph.D. Dissertation, Lehigh University, 1988).

16. N. E. Ryan and J. W. Martin, "The Formation and Stability of Group IVA Carbides and Nitrides in Molybdenum," *J. Less-Common Metals,* 17 (1969) 363-376.

17. A. Kumar and B. L. Eyre, "Grain Boundary Segregation and Intergranular Fracture in Molybdenum," *Proc. R. Soc. Lond. A,* 370 (1980) 431-458.

18. A. Kumar and B. L. Eyre, "An Electron Microscope Study of Mo_2C Precipitates in Molybdenum," *Acta Metall.,* 26(4) (1978) 569-578.

19. W. H. Kearns, ed., *Welding Handbook,* Seventh Edition, Vol. 2 (American Welding Society, Miami, 1978) 138.

20. J. Wadsworth, G. R. Morse, and P. M. Chewey, "The Microstructure and Mechanical Properties of a Welded Molybdenum Alloy," *Mater. Sci. Eng.,* 59 (1983) 257-273.

21. N. E. Ryan, W. A. Soffa, and R. C. Crawford, "Orientation and Habit Plane Relationships for Carbide and Nitride Precipitates in Molybdenum," *Metallography,* 1 (1968) 195-220.

22. J. A. Shields, "Carbide and Nitride Morphology and Stability in Molybdenum Alloys," *Physical Metallurgy and Technology of Molybdenum and Its Alloys,* K. H. Miska, M. Semchyshen, and E. P. Whelan, eds. (AMAX Specialty Metals Corp., Greenwich, CT, 1985) 119-126.

FUSED SALT ELECTROLYSIS OF REFRACTORY METALS

Donald R. Sadoway

Department of Materials Science and Engineering
Massachusetts Institute of Technology
Cambridge, Massachusetts 02139

Abstract

Fused salt electrolysis is used extensively in the primary extraction of metals (electrowinning), in the refining of metals (electrorefining), and in the formation of coatings (electroplating). In the case of refractory metals, fused salt electrolysis competes with nonelectrochemical processes to make the same products. However, fused salt electrolysis has the potential to produce these metals in a variety of technologically useful forms: thin films and epitaxial layers, powders, and a variety of nonequilibrium structures. The theory and practice of fused salt electrolysis as it applies to the extraction, refining, and plating of refractory metals are described. The potential of fused salt electrolysis to synthesize advanced materials containing the refractory metals is assessed.

Refractory Metals: State-of-the-Art 1988
Edited by P. Kumar and R.L. Ammon
The Minerals, Metals & Materials Society, 1989

Introduction

The refractory metals comprise the elements titanium, zirconium, hafnium, vanadium, niobium, tantalum, chromium, molybdenum, and tungsten. They are also known as the transition elements and are found in the Periodic Table in Groups 4, 5, and 6[*]. Although metallurgists find it convenient to refer to the refractory metals as a class, their distinctive electron configurations give each element a character of its own. The exception is the pair, zirconium - hafnium. Owing to the lanthanide contraction, the heavier two elements of each Group are nearly identical in size. However, unlike niobium-tantalum and molybdenum-tungsten, zirconium and hafnium both have identical outer shell electron arrangements, $[Kr]4d^25s^2$ for zirconium and $[Xe]4f^{14}5d^26s^2$ for hafnium, and thus exhibit almost identical chemical behavior. As a result it is rather difficult, for example, to produce hafnium-free zirconium as required in nuclear reactor applications. Appreciation of electronic structure is important in the search for new processes for extraction, purification, and recycling of these metals.

Many of the refractory metals are present in the earth's crust in relatively large proportions (1). For example, titanium, zirconium, chromium, and vanadium rank above zinc, nickel and copper in relative terrestrial abundance. Tungsten is more abundant than tin. Niobium is more abundant than lead. In spite of this, many of the refractory metals are considered strategically important by the United States which, in the case of five of the nine refractory metals, imports in excess of 50 per cent of its requirements (2). With the exception of molybdenum, the refractory metals are extracted primarily from oxygen bearing minerals, e.g., titanium from ilmenite ($FeTiO_2$) and rutile (TiO_2), zirconium from zircon ($ZrSiO_4$), vanadium from a number of vanadates, niobium and tantalum from mixed niobate-tantalate ores, chromium from chromite ($FeO \cdot Cr_2O_3$), and tungsten from tungstates. Molybdenum, in contrast, is primarily extracted from a sulfur bearing mineral, molybdenite (MoS_2). For a description of the metallurgical uses of the refractory metals the reader is directed to literature treating specific elements.

The primary extraction of refractory metals can be accomplished in two ways: thermochemical and electrochemical. One form of thermochemical extraction (practiced especially with Group 4 and 5 elements) involves the metallothermic reduction of a compound containing the refractory metal. The compounds are oxides, fluorides or chlorides, and the reductant is chosen from among aluminum, magnesium, sodium, and calcium, although other reactive metals have been used. Reduction by carbon in an electric furnace is also employed. In another form of thermochemical extraction (practiced mainly with Group 6 elements), the refractory-metal sulfide is roasted to convert it to oxide which is then reduced to metal by hydrogen. Electrochemical extraction involves the electrolytic reduction of a refractory-metal compound dissolved in an ionic medium.

Why fused salt electrolysis? Unfortunately, aqueous solutions are

---

[*] In this article the periodic notation follows recent recommendations by IUPAC and ACS nomenclature committees. To eliminate ambiguity A and B designations are avoided. Groups IA and IIA are denoted 1 and 2, respectively, the d-transition elements groups 3 through 12, and the p-block elements groups 13 through 18. In the last digit the former Roman numeral designation is preserved, e.g., IV → 4 and 14.

unsuitable as electrolytes for the electroreduction of the refractory metals, with the exception of chromium. At decomposition potentials electronegative enough to deposit these metals hydrogen evolution occurs. To a first approximation aqueous electrochemistry is restricted to the energy window bounded by the hydrogen and oxygen evolution reactions. Kinetic factors extend these limitations somewhat, but not far enough to allow electrolysis of the highly reactive refractory metals.

The purpose of this article is to describe the theory and practice of fused salt electrolysis as it applies to the extraction, refining, and plating of refractory metals. As well, the potential of fused salt electrolysis to synthesize advanced materials containing these elements is assessed.

Characteristics of Fused Salt Electrolysis

This author has recently reviewed the fused salt electrolysis of rare earth metals (3). The following account of the characteristics of fused salt electrolysis is substantially identical to that published previously and is reproduced here for the convenience of the reader.

Electrolysis is an electrochemical process, *i.e.*, a process in which chemical reaction is accompanied by electron transfer. Electrolysis is performed in a reactor called an electrochemical cell which is a device that enables electrical energy to do chemical work. A source of the refractory metal is dissolved in a fused salt solution from which metal is extracted by the passage of electrical current.

The principal components of an electrochemical cell are the electrolyte, the electrodes, and the container (sidewalls and floor). The electrolytes are multicomponent melts of either chlorides or fluorides. Each has its advantages and disadvantages. Chlorides offer the advantages of a lower operating temperature and of a greater choice of electrode and container materials. As for disadvantages, chlorides react with moisture: some are hygroscopic, even deliquescent, while others decompose by hydrolysis. At the same time to assure metal purity it is imperative that the electrolyte be free of the impurities formed by reaction with moisture. This adds to the number of unit operations in cell feed preparation from source materials and furthermore puts strict requirements on the operating conditions of the electrolysis cell. For example, if the cell atmosphere becomes contaminated with moisture, it will react with the electrolyte. Among the hydrolysis products are refractory-metal oxychlorides which raise the effective viscosity of the electrolyte and increase anode consumption.

Fluorides, on the other hand, have the advantage of being less reactive with moisture. Additionally, fluorides can dissolve oxides directly. This avoids fluorination which requires reaction with ammonium bifluoride, for example. The use of an oxide-based cell feed in principle simplifies the process flowsheet and reduces capital and operating costs. Unfortunately, the higher melting points and greater corrosivity of the fluorides severely limit the choice of materials of construction in the physical plant. Another disadvantage is the relatively low solubility of refractory-metal oxides in molten fluorides. This limits cell productivity by the need to keep the current density low enough to avoid the onset of anode effect.

During electrolysis electric current is passed from the anode through the electrolyte to the cathode. The electrolyte must be strictly an ionic conductor, while the electrodes must be electronic conductors. The anode can be either consumable or nonconsumable. Consumable anodes are found in various electrolytic processes. In electrorefining, the anode consists of an impure form of the metal to be purified. In electroplating, the consumable anode may act as the feedstock for the deposited metal. In electrowinning, the consumable anode consists of a material that reacts with the products of the anodic reaction, *e.g.*, a carbon anode in the presence of evolving anodic oxygen. Used primarily in electrowinning, nonconsumable anodes consist of materials inert to chemical and electrochemical attack. In chloride electrolytes carbon commonly serves as a nonconsumable anode. There is no fully satisfactory nonconsumable anode for use with fluoride-based electrolytes. Recently, with reference to the Hall cell for the electrolytic extraction of aluminum a new approach to discovering nonconsumable anodes for use in fused salt electrolysis operations has been described along with the relevant set of selection criteria (4).

As for cathode materials, carbon, refractory metals such as tungsten, molybdenum, and tantalum, as well as low carbon steel have all been employed. In practice, the cathode functions only as the current lead. During electrolysis the cathode material is separated from the electrolyte by the product refractory metal which is produced in solid form unless a host low melting metal is employed as a liquid cathode. Production of solid metal and liquid alloy each has its advantages and disadvantages. When the cell product is solid metal it is not necessary to purify it of large amounts of host cathode metal. The disadvantage is that solid metal deposited from molten salts is invariably dendritic. This results in salt entrainment with the need for some form of subsequent treatment for salt removal. The production of liquid metal solves the morphological problem and facilitates easy removal of product from the cell by syphoning; however, this approach has its faults, the principal one being the need to refine the refractory metal out of the product cathode metal alloy.

The electrolyte is contained in a steel shell lined with carbon block. With chlorides a lining of ceramic or low carbon steel is acceptable. It is also possible to operate with a sidewall of frozen electrolyte. Cell voltages depend upon the particular melt chemistry and cell design which determines the contribution of the ohmic resistance of the bath. Current densities are in the neighborhood of 1 A cm^{-2}. Cell operating temperatures span 400° to 1000°C.

Cell productivity is expressed in terms of several figures of merit. Current efficiency can be loosely defined as the ratio of the number of equivalents of metal product to the number of moles of electrical charge delivered to the cell by the power supply. As such, current efficiency is effectively a measure of compliance with Faraday's laws of electrolysis. In electrowinning cells cathodic current efficiencies of 60 to 80% are not uncommon. These figures are well below those reported for the electrolytic production of aluminum which typically attains current efficiencies exceeding 90%. Voltage efficiency is the ratio of the equilibrium decomposition potential to the applied cell voltage. Quite simply, voltage efficiency expresses the deviation from the Nernst equation and is a measure of inefficiency due to kinetic factors, *e.g.*, the physical resistance of the electrolyte, the electrodes, bus bars, and contacts as well as overvoltages associated with the faradaic processes occurring at each electrode. In the electrolysis of fluorides voltage efficiencies are typically below 50%, while in the electrolysis of chlorides values below 25% have been reported.

Fused salt electrolysis can also generate coatings containing refractory metals. The range of compositions includes elements, alloys, and compounds. In a recent review of fused salt electroplating the features of this process were described, and the literature from 1967 through 1986 was surveyed (5). That review broadly treated the entire field of fused salt electroplating and cited refractory metals only in passing. This article focuses on refractory metals.

Electrowinning and Electrorefining

For the purposes of this brief review, advances reported in the literature since 1978 are cited. For an accounting up to that time the reader is directed to earlier reviews such as that by Inman and White (6).

With the exception of titanium, the primary extraction of refractory metals by fused salt electrolysis has received limited attention. Accounts of research on this topic include the following: titanium (7-10), zirconium (11, 12), vanadium (13), and in general (14). Preparation of high purity metal and recovery of scrap, both by fused salt electrorefining, have been under study and are described in these reports: titanium (15-17), zirconium (18, 19), niobium (20, 21), and Group 4 and 5 metals (22).

Along with the extraction and refining processes themselves, various aspects of molten salt chemistry and electrochemistry have been under investigation. Topics include electrolyte optimization (23, 24), anodic reaction (25), chronopotentiometry (26), disproportionation (27), voltammetry (28, 29), exchange current density measurements (30), phase diagram determinations (31), and vapor pressure measurements (32). Laboratory scale electrolysis cells have been operated to investigate how process parameters affect current efficiency (33). Finally, attempts have been made to construct mathematical models of the electrolysis process (34).

Electroplating

Fused salt electroplating of refractory metals has received somewhat more attention than extraction. At least on the basis of reports in the open literature, the most thoroughly studied metals from the perspective of electroplating are molybdenum (35-38) and titanium (39, 40). Other metals, however, continue to receive attention: niobium (41, 42), tantalum (41, 43, 44), chromium (45-48), and tungsten (49). Electrodeposited coatings can be pure metal, metal alloy, or compounds. Reports of research on the electrodeposition of alloys include Mo-Nb (50), Ni-Ta (51) and Ni-Nb (51). As for electrodeposition of compounds containing refractory metals, the earliest work was concentrated on the production of carbides, largely for the purposes of generating abrasion resistant coatings (49, 52). More recently, attention has turned to a broader array of materials syntheses, including metal matrix composites (53) and refractory-metal oxides containing controlled subvalent forms of the refractory-metal ion (54). In the case of the latter, electrosynthesis from fused salts has facilitated the generation of compounds otherwise produced thermochemically at high temperatures and high pressures. An alternative to electrosynthesis of refractory-metal oxide compounds is anodization. The distinction between the two processes is that anodization is a process for the faradaic oxidation of elemental metal present as the anode substrate. Electrosynthesis is a process for the faradaic discharge of both metal and oxygen (both present in the electrolyte) to form a compound on an electrode substrate. While refractory metals can be anodized in aqueous media, fused salt anodization offers certain advantages in product quality, both in terms of chemical composition and microstructure (55).

159

Future Directions

When viewed in the context of materials processing, fused salt electrolysis has many fine attributes. However, unresolved technical issues continue to impede its further commercialization. These include low cathode current efficiency, anode effect (restricted to fluoride-based electrolytes processing oxide feed), purity of metal product, corrosion of cell components, and heat balance of the cell. Losses of power efficiency can be traced to the fact that the refractory metals exhibit multiple valency in these melts. This can lead to redox looping or parasitic reaction of subvalent ions with metal product. Furthermore, as mentioned above, most cells produce solid metal. Solid electrodeposits obtained in molten salts are typically dendritic or powdery, but rarely smooth, especially when the metal is of high purity. To deal with multiple valency one can optimize bath chemistry (23, 24), use a diaphragm to separate specified ions from one another while allowing selected mass transfer between regions of the cell, an approach very popular with designers of titanium electrowinning cells (56-58), or invent divided cells featuring staged reactions (59). The morphological problem has been attacked in some cases by employing molten metal cathodes (for example, 9, 14). The use of leveling agents, while potentially beneficial, has been examined for electroplating (35), but evidently not for electrowinning where the goal is to produce metal of the highest purity.

Besides the strictly technical issues there is a host of other concerns. For example, in designing new electrolytic processes one must look carefully at capital costs as well as operating costs. If fused salt electrolysis is to compete effectively with nonelectrochemical extraction processes, steps must be taken to design flowsheets with a minimum number of unit operations. More attention must be paid to environmental, health and safety issues. Care must be exercised to ensure that the production of refractory metal not be accompanied by the generation of any form of hazardous or toxic waste.

From the perspective of a technology capable of generating advanced materials, fused salt electrolysis looks extremely attractive (5). Again, its full potential has hardly been exploited. Advanced materials are characterized not only by their specialized chemistries, i.e., purity, doping level, etc., but also by their tailored microstructures, which fused salt electrolysis has the potential to generate. These include thin films, epitaxial layers, powders, and various nonequilibrium structures. Powders of a number of refractory metals have been produced by fused salt electrolysis: titanium (60-62), niobium (63), tantalum (63), chromium (61), tungsten (61, 64), and in general (65). As for nonequilibrium structures, fused salt electrolysis has the capacity to produce metastable phases, compositionally graded microstructures, and compositionally modulated microstructures. To date there has been very little use of fused salt electrolysis to generate nonequilibrium structures of refractory metals. There are isolated reports in the literature of metastable phase formation (66) and deposition of single crystals with preferred orientation (67), but overall, this field is severely underexploited.

In summary, fused salt electrolysis is a viable technology for producing refractory metals along with their alloys and compounds in a variety of technologically important forms. Some scientific and technical problems remain to be solved before the full potential of this technology will be felt in the marketplace.

Acknowledgement

The author gratefully acknowledges the assistance of Ms. Heather Brooks Shapiro whose knowledge and understanding of commands, indexing terms, and search tools facilitated the computer assisted review of the literature.

References

1. Van Nostrand's Scientific Encyclopedia, 7th edition, ed. D.M. Considine (New York NY: Van Nostrand Reinhold, 1989), 578.

2. ibid., 579.

3. D.R. Sadoway, "Fused Salt Electrolysis of Rare Earth Metals," in Rare Earths, ed. R.G. Bautista and M.M. Wong (Warrendale PA: TMS-AIME, 1988), 345-353.

4. A.D. McLeod, J.-M. Lihrmann, J.S. Haggerty, and D.R. Sadoway, "Selection and Testing of Inert Anode Materials for Hall Cells," Light Metals 1987, ed. R.D. Zabreznik (Warrendale PA: TMS-AIME, 1987), 357-365.

5. D.R. Sadoway, "Fused Salt Electroplating," Electrodeposition Technology, Theory and Practice, ECS Symposium Vol. 87-17, ed. L.T. Romankiw and D.R. Turner (Pennington NJ: The Electrochemical Society, 1987), 397-413.

6. D. Inman and S.H. White, "The production of refractory metals by the electrolysis of molten salts; design factors and limitations," J. Appl. Electrochem., 8 (5) (1978), 375-390.

7. P.R. Juckniess and D.R. Johnson, inventors, Dow Chemical Co., assignee, "Apparatus for electrowinning multivalent metals," U.S. patent, no. 4,116,801, Sept. 26, 1978.

8. G. Lorthioir and M. Nardin, inventors, Agence Nationale de Valorisation de la Recherche, assignee, "Titanium by electrolytic reduction in a bath of molten titanium halides," French patent, no. 2,359,221, Feb. 17, 1978.

9. M. Onozawa, inventor, Nippon Steel Corp., assignee, "Preparation of titanium or its alloy by fused salt electrolysis," Japanese patent, no. 63/118089 A2 [88/118089], May 23, 1988.

10. J. Cohen and G. Lorthioir, inventors, Pechiney S.A., assignee, "Method and apparatus for electrodeposition of a metal from a molten halide salt bath," French patent, no. 2,560,896, Sept. 13, 1985.

11. P. Pint and S.N. Flengas, "Production of zirconium metal by fused salt electrolysis," Trans. Inst. Min. Metall., 87 C (March) (1978), 29-49.

12. A.V. Kovalevskii, L.V. Kovalevskaya, and I.F. Nichkov, "Electrolysis conditions for $KCl-K_2ZrF_6$ and $NaCl-K_2ZrF_6$ molten salt mixtures," Tsvetn. Met., 1983, no. 8: 62-64.

13. L.E. Ivanovskii, A.V. Lukinskikh, and V.P. Batukhtin, "Production of refractory metals by electrolysis of molten halide baths," deposited document, VINITI 502-79, 45-8 (1979); Chem. Abs. 92:118408s.

14. G.F. Warren, A. Horstik, A. Corbetta, R.E. Malpas, A. Honders, and G.J. Van Eijden, inventors, Shell Internationale Research Maatschappij B.V., assignee, "Process for the electrolytic production of metals," European patent appl., no. 219,157 A1, Apr. 22, 1987.

15. E. Tanaka, T. Kikuchi, and T. Tsumori, inventors, Sony Corp., assignee, "Reduction of high valent titanium in a fused salt bath used for titanium refining," Japanese patent, no. 53/22111 [78/22111], Mar. 1, 1978.

16. K. Shimotori, Y. Ochi, H. Ishihara, T. Umeki, and T. Ishigami, inventors, Toshiba Corp., assignee, "Highly pure titanium and process for producing it," European patent appl., no. 248,338 A1, Dec. 9, 1987.

17. J.M.J. Paixao, F. Teixeira de Almeida, and R.L. Combes¡, inventors, Companhia Vale do Rio Doce, assignee, "Electrolytic purification of titanium obtained by aluminothermal or magnesium-thermal reduction of anatase concentrates," Brazilian patent, no. 84/5093 A, May 13, 1986.

18. S.N. Chintamani, P. Pande, A.K. Taneja, and J.C. Sehra, "Recycling off-grade Zircaloy scrap using a molten salt refining process," ASTM Spec. Tech. Publ., 939 (Zirconium Nucl. Ind.) (1978), 136-145.

19. G.J. Kipouros and S.N. Flengas, "Electrorefining of zirconium metal in alkali chloride and alkali fluoride fused electrolytes," J. Electrochem. Soc., 132 (5) (1985), 1087-1098.

20. A.P. Khramov, L.E. Ivanovskii, and V.P. Batukhtin, "Effect of vibration on cathode polarization and current efficiency in refining of niobium in chloride-fluoride melts," Elektrokhimiya, 21 (6) (1985), 802-804.

21. K. Schulze and M. Krehl, "The preparation of pure niobium for neutron dosimetry purposes," Nucl. Instrum. Methods Phys. Res., A 236 (3) (1985), 609-616.

22. M. Armand and J.P. Garnier, inventors, Pechiney S.A., assignee, "Method for improving the purity of transition metals obtained by electrolysis of their halides in molten salt baths," French patent, no. 2,579,230 A1, Sept. 26, 1986.

23. A.L. Glagolevskaya, S.A. Kuznetsov, E.G. Polyakov, and P.T. Stangrit, "Effect of anion composition of electrolytes on the disproportionation reactions of transition metal compounds in molten salts," Rasplavy, 1 (6) (1987), 81-85.

24. A.V. Kovalenskii, A.E. Mordovin, I.F. Nichkov, V.I. Shishalov, and A.V. Kovalevskii, "Physicochemical properties of molten NaCl - Na_2ZrF_6 mixtures," Izv. Vyssh. Uchebn. Zaved., Tsvetn. Metall., no. 5: 1987, 115-116.

25. G.S. Chen, M. Okido, and T. Oki, "Electrochemical studies of titanium in fluoride-chloride molten salts," J. Appl. Electrochem., 18 (1) (1988), 80-85.

26. C.A.C. Sequeira, "Chronopotentiometric study of titanium in molten NaCl + KCl + K_2TiF_6," <u>J. Electroanal. Chem. Interfacial Electrochem.</u>, 239 (1-2) (1988), 203-208.

27. V.I. Shapoval, V.I. Taranenko, and I.V. Zarutskii, "Electrochemical behavior of the titanium(III)/titanium(II) system in molten chlorides," <u>Ukr. Khim. Zh.</u>, 53 (4) (1987), 370-374.

28. Z. Qiao and P. Taxil, "Electrochemical reduction of niobium ions in molten LiF - NaF," <u>J. Appl. Electrochem.</u>, 15 (2) (1985), 259-265.

29. L.P. Polyakova, B.I. Kosilo, E.G. Polyakov, and A.B. Smirnov, "Electrochemical behavior of tantalum in a CsCl - KCl - NaCl - $TaCl_5$ melt," <u>Elektrokhimiya</u>, 24 (7) (1988), 892-897.

30. J.L. Settle and Z. Nagy, "Metal deposition-dissolution in molten halides: on the question of measurability of very fast electrode reaction rates," <u>J. Electrochem. Soc.</u>, 132 (7) (1985), 1619-1627.

31. K. Koyama and Y. Hashimoto, "Liquidus surfaces of the KF - B_2O_3 - Li_2WO_4 - Na_2WO_4 - K_2WO_4 systems," <u>J. Less-Common Met.</u>, 141 (1) (1988), 55-58.

32. M.V. Smirnov, A.B. Salyulev, and V.Ya. Khudyakov, "Effect of the ionic composition of electrolytes on volatility of $HfCl_4$ and the potential of metal deposition on a cathode," <u>Fiz. Khim. Elektrokhim. Redk. Met. Solevykh Rasplavakh</u>, P.T. Stangrit, editor, (Apatity, USSR: Akad. Nauk SSSR, Kol'sk. Fil., 1984), 3-8.

33. A.V. Kovalevskii and V.V. Toropov, "Current efficiency in the electrolysis of NaCl - K_2ZrF_6 and NaCl - KCl - K_2ZrF_6 molten salt mixtures," <u>Izv. Vyssh. Uchebn. Zaved., Tsvetn. Metall.</u>, no. 5: 1984, 51-55.

34. S.L. Gol'dshtein, S.V. Gudkov, S.P. Raspopin, G.B. Smirnov, and P.Z. Saifullin, " Experimental static model of potentiostatic cathodic deposition of titanium from a chloride melt," <u>Izv. Vyssh. Uchebn. Zaved., Tsvetn. Metall.</u>, no. 2: 1985, 31-36.

35. G.J. Kipouros and D.R. Sadoway, "The electrodeposition of improved molybdenum coatings from molten salts by the use of electrolyte additives," <u>J. Appl. Electrochem.</u>, 18 (6) (1988), 823-830.

36. V.I. Shapoval, A.N. Baraboshkin, Kh.B. Kushkhov, and V.V. Malyshev, "Specific features of the electroreduction of MoO_3 forms in the presence of a tungstate melt," <u>Elektrokhimiya</u>, 23 (7) (1987), 942-946.

37. K. Koyama, Y. Hashimoto, and K. Terawaki, "Smooth electrodeposits of molybdenum from KF - $K_2B_4O_7$ - K_2MoO_4 fused salt melts," <u>J. Less-Common Met.</u>, 134 (1) (1987), 141-151.

38. T. Hatusika, M. Miyake, and T. Suzuki, "Formation of refractory metal films in low temperature molten salt bath," <u>Kenkyu Hokoku - Asahi Garasu Kogyo Gijutsu Shoreikai</u>, 49 (1986), 289-293.

39. X. Gu, S. Duan, and D. Inman, "Electroplating of titanium in molten LiCl - KCl eutectic containing lower-valent titanium," <u>Xiyou Jinshu</u>, 7 (3) (1988), 182-186.

40. A. Robin, J. De Lepinay, and M.J. Barbier, "Electrolytic coating of titanium onto iron and nickel electrodes in the molten LiF + NaF + KF eutectic," J. Electroanal. Chem. Interfacial Electrochem., 230 (1-2) (1987), 125-141.

41. P. Taxil and J. Mahenc, "Formation of corrosion-resistant layers by electrodeposition of refractory metals or by alloy electrowinning in molten fluorides," J. Appl. Electrochem., 17 (2) (1987), 261-269.

42. G.P. Capsimalis, E.S. Chen, R.E. Peterson, and I. Ahmad, "On the electrodeposition and characterization of niobium from fused fluoride electrolytes," J. Appl. Electrochem., 17 (2) (1987), 253-260.

43. A.W. Berger, "Fused-salt electrodeposited tantalum coatings," Chem.-Anlagen Verfahren, no. 3: 1980, 82-84.

44. P. Los, J. Josiak, A. Bogacz, and W. Szklarski, "Tantalum coatings deposition from fluoride electrolytes," Arch. Hutn., 29 (4) (1984), 515-527.

45. T. Vargas, R. Varma, and A. Brown, "Electrodeposition of microcrystalline chromium from fused salts," Molten Salts, ECS Symposium Vol. 87-7, ed. G. Mamantov, M. Blander, C. Hussey, C. Mamantov, M.-L. Saboungi, and J. Wilkes (Pennington NJ: The Electrochemical Society, 1987), 1018-1027.

46. A.M. Emsley and M.P. Hill, "The corrosion and deposition performance of molten salt electrodeposited chromium coatings," J. Appl. Electrochem., 17 (2) (1987), 283-293.

47. T. Vargas and D. Inman, "Controlled nucleation and growth in chromium electroplating from LiCl - KCl melt," J. Appl. Electrochem., 17 (2) (1987), 270-282.

48. R.A. Bailey and T. Yoko, "High-temperature electroplating of chromium from molten FLINAK," J. Appl. Electrochem., 16 (5) (1986), 737-744.

49. H. Yabe, Y. Ito, K. Ema, and J. Oishi, "Electrodeposition of tungsten and tungsten carbide from molten halide ," Molten Salts, ECS Symposium Vol. 87-7, ed. G. Mamantov, M. Blander, C. Hussey, C. Mamantov, M.-L. Saboungi, and J. Wilkes (Pennington NJ: The Electrochemical Society, 1987), 804-813.

50. Z.I. Valeev, A.N. Baraboshkin, Z.S. Martem'yanova, and N.O. Esina, "Electrodeposition of molybdenum -niobium alloys from their chloride melt," Elektrokhimiya, 24 (1) (1988), 59-63.

51. Z. Qiao and P. Taxil, "Electrochemical surface alloying on nickel with tantalum and niobium in molten fluorides and properties of tantalum - nickel and niobium - nickel alloys," Jinshu Xuebao, 23 (2) (1987), B76-B83.

52. K.H. Stern and S.T. Gadomski, "Electrodeposition of tantalum carbide coatings from molten salts," J. Electrochem. Soc., 130 (2) (1983), 300-305.

53. M.E. De Roy and J.P. Besse, "Synthesis of single-crystal inorganic compounds by electrochemical reduction in a molten medium," Rev. Int. Hautes Temp. Refract., 24 (2) (1987), 71-83.

54. G.A. Hope and R. Varma, "Molten salt deposition of metal matrix composite materials," *Aust. J. Chem.*, 41 (8) (1988), 1257-1259.

55. V.P. Yurkinskii, E.G. Firsova, A.G. Morachevskii, and A.A. Maiorov, "Mechanism and kinetics of the electrochemical oxidation of molybdenum and tungsten in molten alkali metal nitrates," *Zh. Prikl. Khim. (Leningrad)*, 57 (3) (1984), 695-698.

56. Sony Corp., assignee, "Diaphragm for fused salt electrolysis," Japanese patent, no. 56/5832 [81/5832], Feb. 6, 1981.

57. E. Chassaing, F. Basile, and G. Lorthioir, "Fused-salt electrolysis for production of titanium metal - present state and future developments," *Titanium '80 Science and Technology*, ed. H. Kimura and O. Izumi (Warrendale PA: TMS-AIME, 1980), 1963-1967.

58. G. Cobel, J. Fisher, and L.E. Snyder, "Electrowinning of titanium from titanium tetrachloride: pilot plant experience and production plant projections," *Titanium '80 Science and Technology*, ed. H. Kimura and O. Izumi (Warrendale PA: TMS-AIME, 1980), 1969-1976.

59. M.V. Ginatta, private communication, "Industrial plant for the production of electrolytic titanium, Ginatta technology," report RT 88-03-077, Ginatta S.A., Torino, Italy, March 1988.

60. G.P. Dovgaya, V.V. Nerubashchenko, S.P. Chernysheva, and L.K. Mineeva, "Effect of the electrolyte composition on the properties of electrolytic titanium powders," *Poroshk. Metall. (Kiev)*, no. 10: 1987, 6-10.

61. A.B. Suchkov, A.S. Vorob'eva, V.N. Kryzhova, L.V. Ryumina, A.G. Kaganov, I.V. Chikunova, and B.F. Kovalev, "Effect of electrolyte composition and current density on the particle size of electrolytic powders," *Poroshk. Metall. (Kiev)*, no. 6: 1987, 1-4.

62. S.L. Gol'dshtein, S.V. Gudkov, S.P. Raspopin, and G.B. Smirnov, "Effect of concentration on the formation of fine titanium powders by potentiostatic deposition from chloride melts," *Izv. Vyssh. Uchebn. Zaved., Tsvetn. Metall.*, no. 2: 1986, 58-61.

63. C.F. Rerat, inventor, Fansteel, Inc., assignee, "Tantalum and niobium powder," U.S. patent, no. 4,149,876, Apr. 17, 1979.

64. V.A. Pavlovskii and V.A. Reznichenko, "Electrolytic method for manufacturing coarse tungsten powder," *Poroshk. Metall. (Kiev)*, no. 11: 1986, 1-3.

65. M. Armand, "Process for elaboration of transition metal powders in molten salt baths," French patent, no. 2,592,664 A1, July 10, 1987.

66. K.A. Kaliev, A.N. Baraboshkin, and S.M. Zakhar'yash, "Formation of metastable phases during the electrolysis of Na_2WO_4 - Li_2WO_4 - WO_3 system melts," *Elektrokhimiya*, 20 (3) (1984), 328-331.

67. A.N. Baraboshkin, Z.S. Martem'yanova, S.V. Plaksin, and N.O. Esina, "Preferred orientation of the growth of metals electroplated from fused salts. Relation between crystal habit and direction of the orientation axis," *Elektrokhimiya*, 14 (1) (1978), 9-15.

SUPERCOOLING EFFECTS IN NB-RICH NB-SI ALLOYS

M.D. Lipschutz, A.B. Gokhale, and G.J. Abbaschian

Department of Materials Science and Engineering
University of Florida
Gainesville, FL 32611

Abstract

Bulk Nb-Si alloys in the range 15 to 22 at.% Si were supercooled and rapidly solidified via electromagnetic levitation processing. Microstructures of the processed samples were studied by compositional, morphological, and x-ray structural analyses. Based on these results, the zone of coupled growth for the (Nb)+Nb_3Si eutectic was deduced. The solidification sequence in the processed alloys is described as a function of the degree of melt supercooling, and the shape and extent of the coupled zone. It is shown that bulk supercooling leads to the formation of metastable (Nb)+Nb_5Si_3 eutectic, metastable Nb_4Si, and amorphous phases.

Refractory Metals: State-of-the-Art 1988
Edited by P. Kumar and R.L. Ammon
The Minerals, Metals & Materials Society, 1989

Introduction

Nb-Si alloys are of considerable interest for high temperature structural applications because silicon has the potential to provide increased strength through solid solution or dispersion strengthening mechanisms. In addition, Si offers the possibility of improved high temperature oxidation resistance through the formation of a protective oxide layer. However, Si amounts greater than 5 at.% can lead to a loss of ductility due to the presence of brittle intermediate phases. Thus, current research is focused on improving the ductility of Nb-Si alloys through the formation of nano-scale microstructures or replacing the brittle stable phases with more ductile metastable ones.

In earlier studies of Si-rich Nb-Si alloys [1,2], various microstructural consequences of rapid solidification processing (RSP) via bulk supercooling were investigated. It was shown that the nucleation hierarchy of various phases in the supercooled liquid had a significant effect on the microstructures. For example, in samples where a primary phase nucleation of Nb_5Si_3 was observed during supercooling, the resultant microstructures were extremely heterogeneous both morphologically and compositionally. On the other hand, when the primary phase nucleation was suppressed, the microstructures exhibited only a fine cellular eutectic with a quasi-regular morphology. Upon nucleation and growth in the supercooled melt, the intermetallic Nb_5Si_3 was found to contain Si approximately 4% in excess of the equilibrium value. This metastable supersaturation was shown to aid in the subsequent solid state decomposition of the phase.

It was further shown that the solidification sequence in the supercooled alloys could be deduced by considering factors such as: original alloy composition, degree of bulk supercooling, potency of heterogeneous nucleants and the nucleation hierarchy of various phases, the shifts in the composition and temperature of the liquid upon primary phase nucleation and recalescence, the shape and extent of the zone of coupled eutectic growth, and the external cooling conditions.

In this paper, the compositional and morphological effects associated with bulk melt supercooling and rapid solidification of Nb-rich Nb-Si alloys are described. The determination of the coupled eutectic zone through analysis of the primary solidification morphology as a function of composition and melt supercooling will be presented. In addition, it will be shown that these analyses can be effectively used to deduce the solidification sequence in the bulk supercooled alloys. Finally, metastable phase formation in alloys ranging from 17 to 20 at.% Si will be described.

The Nb-Si binary phase diagram is shown in Figure 1, together with the range of compositions investigated (dashed lines). The compositions shown on the (Nb)+ϵ eutectic isotherm were measured during this investigation via microprobe analysis. A selected portion of the equilibrium diagram (Nb to Nb_5Si_3), together with the calculated metastable extensions is shown in Figure 2. Particular attention is drawn to the metastable (Nb)+γ (Nb_5Si_3) eutectic which occurs upon suppression of ϵ.

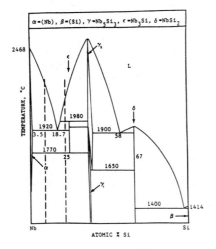

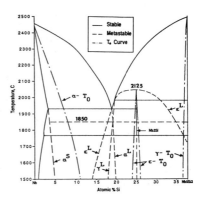

Figure 1 - Nb-Si equilibrium phase diagram. Dashed lines indicate range of alloy compositions studied.

Figure 2 - Nb to Nb_5Si_3 portion of the phase diagram indicating calculated metastable constructions.

The coordinates of the eutectic were calculated to be 1850°C and 19 at.% Si. Also, note that upon γ suppression, the intermediate phase ε becomes a congruently melting compound with a melting temperature of 2025°C and does not undergo the eutectodial decomposition.

Experimental Procedure

Nb-Si alloys ranging from 15 to 22 at.% Si were prepared from the pure components by arc melting 1.0 to 1.5 gram buttons under a purified Ar atmosphere. The purities of Nb and Si were 99.99 at.% and 99.999 at.%, respectively. The arc melted buttons were processed by EM levitation in a Ti-getter purified atmosphere of Ar and He. The experimental set-up, as described in detail elsewhere [3-6], is illustrated schematically in Figure 3. Temperature measurements of the levitated samples were made

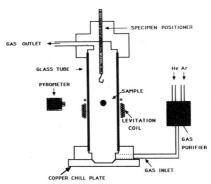

Figure 3 - Schematic of the EM levitation system

using a single color pyrometer calibrated using a known thermal arrest, e.g., the eutectic isotherm. The processing involved bulk supercooling of the molten alloys, followed by rapid quenching of the supercooled melt. Quenching media included a copper chill plate or a copper-platen splat quenching apparatus. The processed samples were analyzed by electron microprobe, SEM, TEM, x-ray powder diffraction, and optical microscopy to obtain compositional, morphological, and structural information. In all instances, the reported supercoolings are with respect to the (Nb)+ϵ (Nb$_3$Si) eutectic isotherm (1920°C).

Results and Discussion

Various types of metastable structures were observed in alloys solidified from a supercooled state. In describing these structures and understanding their formation mechanisms, it is important to distinguish between structures which are morphologically and/or topologically metastable and those which are compositionally metastable [7], although the two are not mutually exclusive. Moreover, it is also possible to form metastable phases which have crystalline structures different from the stable phases.

Examples of topologically or morphologically metastable structures include high-internal-energy nano-scale eutectics [8], eutectic morphologies from off-eutectic compositions [9], etc. As discussed below, these morphologies may be understood in terms of kinetic selection criteria for various growth forms and the local growth conditions at the solid/liquid interface. Microstructures with such metastable morphologies may then include phases which are themselves metastable at the solidification temperature. Consequently, the results presented below are divided into two sections: (A) Coupled eutectic growth and (B) Metastable phase formation.

(A)   Coupled Eutectic Growth

The coupled zone delineates the temperature-composition limits in which the eutectic morphology is the kinetically preferred growth form. In this study, the zone of coupled growth for the (Nb)+ϵ eutectic was determined experimentally by quenching alloys in the range 15 to 22 at.% Si from various degrees of bulk supercooling. For a given bulk supercooling, an alloy was considered to lie inside the coupled zone if only the coupled eutectic morphology was observed at the quenched surface (i.e., the surface in contact with the Cu-chill). On the other hand, alloys exhibiting a primary phase at the quenched surface were considered to lie outside the coupled zone.

The coupled zone thus determined is shown in Figure 4. It can be seen that the coupled zone is skewed toward the ϵ side, which is indicative of kinetic attachment difficulties during the growth of this phase. This is supported by our thermodynamic calculations which yielded a value of approximately 4.5 for the dimensionless entropy of fusion ($\Delta S_f/R$) for ϵ, and our experimental observations, as illustrated in Figure 5.

As mentioned earlier, the shape and extent of the coupled zone is a consequence of competition between various growth forms. Thus, for a given interfacial supercooling, the growth form with the highest growth velocity (or for a given growth velocity, the growth form with the lowest interfacial supercooling) "leads" the growth front. This is illustrated

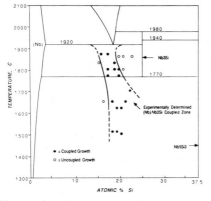

Figure 4 - Experimentally determined
Nb-(Nb)+ϵ coupled zone.

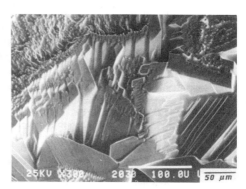

Figure 5 - SE image of
faceted ϵ in a 20 at.% Si
alloy.

in Figures 6(a) to 6(c) where the interfacial supercoolings for (Nb),
(Nb)+ϵ eutectic, and ϵ are plotted as a function of growth velocity for
three compositions. These curves were calculated based on the dendritic
and eutectic growth models proposed by Kurz and Fisher [10]; the crossing
points of the various curves correspond to the transition between the
growth morphologies.

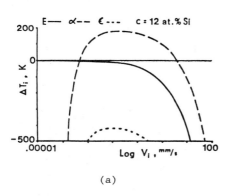

(a)

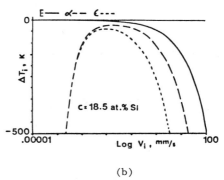

(b)

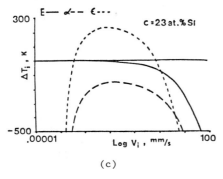

(c)

Figure 6 - Growth velocity vs.
interfacial supercooling for:
(a) Nb-12 at.% Si, (b) Nb-18.5
at.% Si, and (c) Nb-23 at.% Si.

171

Figure 6(a) indicates that for a 12 at.% Si alloy the eutectic is stable at small levels of supercooling, but as the supercooling is increased there is a transition to (Nb) dendrites leading the growth front. Figure 6(b) shows that for an 18.5 at.% Si alloy, the eutectic is stable over the entire range of supercoolings. On the other hand, for the Nb- 23 at.% Si alloy, Figure 6(c), there are two crossing points between the eutectic and ϵ curves.

This behavior is apparent in the skewing of the coupled zone toward ϵ, which results in a morphological sequence of eutectic $\rightarrow$ ϵ $\rightarrow$ eutectic with increasing interfacial supercooling for this hypereutectic alloy. It should be noted that a coupled zone will also exist for the metastable (Nb)+γ eutectic. However, due to the infrequent appearance of this eutectic in the observed microstructures, it was not possible to determine either the shape or the extent of this zone experimentally.

The determination of the shape and extent of the coupled zone is critical to the understanding of solidification sequences in supercooled alloys. This is illustrated below with three examples:

<u>(1) Suppression of Primary (Nb) in Nb-15 at.% Si.</u>   The microstructure of a Nb-15 at.% Si (hypoeutectic) sample supercooled 80 K and quenched on a Cu-chill plate is shown in Figure 7(a). The back scattered electron micrograph shows a coupled (Nb)+ϵ eutectic which forms at the quenched surface at the beginning of solidification. It can be seen from Figure 4 that, at this supercooling, the (Nb)+ϵ eutectic is the kinetically favored growth morphology. However, as a result of the release of heat of fusion during eutectic growth, the temperature of the remaining liquid rises above the eutectic isotherm, where growth of the coupled eutectic stops. The primary (Nb) dendrites then continue growing on the (Nb) lamellae in the (Nb)+ϵ eutectic, as shown by the higher magnification micrograph in Figure 7(b). Note that for this composition, if the initial supercooling were greater than ~ 100 K, the primary growth morphology would be (Nb) dendrites (See Figure 4).

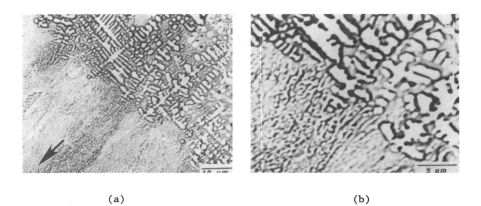

(a)   (b)

Figure 7 - BSE micrographs of Nb-15 at.% Si supercooled 80 K and quenched on Cu-chill. (a) growth of coupled (Nb)+ϵ eutectic near the chill surface (chill direction indicated by arrow) with transition to (Nb) dendrites leading the growth front, and (b) growth of (Nb) dendrites from the existing (Nb) lamellae in the eutectic.

The microstructure formed during initial stages of solidification may be classified as compositionally and morphologically metastable: the coupled eutectic has an average composition equal to the initial alloy composition (15 at.% Si), rather than the equilibrium concentration of 18.7 at.%. Morphologically, under normal solidification, the alloy should consist of primary (Nb) dendrites with an interdendritic (Nb)+ε eutectic containing 18.7 at.% Si.

(2) Suppression of ε in Nb-17 at.% Si. The microstructure of the alloy solidified in a "normal" manner consists of primary (Nb) dendrites surrounded by (Nb)+ε eutectic containing approximately 18.7 at.% Si, as shown in Figure 8. Microstructures of the alloy quenched on a Cu-chill plate from a supercooling of 200 K are shown in Figures 9(a) and 9(b) and Figures 10(a) and 10(b). Figure 9(a) shows the microstructure near the chill surface while Figure 9(b) shows the corresponding compositional analysis. The microstructure in the sample interior (i.e., away from the chill surface) is illustrated in Figure 10(a), with the corresponding compositional analysis shown in Figure 10(b).

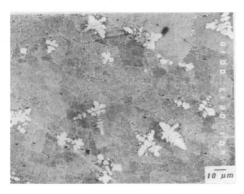

Figure 8 - BSE image of Nb-17 at.% Si quenched from superheat of 90 K (white = (Nb), matrix = (Nb)+ε eutectic).

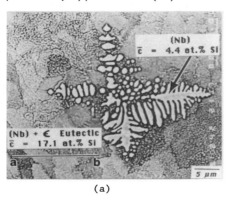

(a)

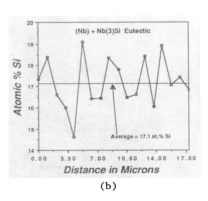

(b)

Figure 9 - Nb-17 at.% Si supercooled 190 K and splat-quenched: (a) BSE image of supersaturated (Nb) in (Nb)+ε eutectic matrix, and (b) corresponding linescan in the matrix.

As shown in Figure 9(a), the microstructure near the chill surface consists of primary (Nb) dendrites surrounded by (Nb)+ε eutectic. The composition of the primary (Nb) dendrites was found to be 4.4 at.% Si (corresponding to an increased solubility of ~1 at.% Si), while the compositional linescan in Figure 9(b) (line a-b in Figure 9(a)) indicates an average eutectic composition of 17.1 at.% Si, i.e., slightly in excess of the original alloy composition. Note that on the linescan, none of the compositional peaks and valleys correspond to the single phase compositions because the eutectic spacing is smaller than 1 μm, i.e., the approximate resolution limit of the microprobe.

The microstructure in the interior (Figure 10(a)) is somewhat more complex due to the presence of γ (black phase in the back scattered micrograph). The identification of this phase was confirmed by microprobe analysis which yielded a composition of 34 at.% Si. It should be noted that although this composition does not correspond exactly to stoichiometric Nb₅Si₃, the analysis may have been affected by surrounding areas due to the small scale (~ 1 μm) of this phase. The analysis, however, clearly indicates that the phase is not Nb_3Si.

The presence of γ in this sample is interesting because of the two different morphologies in which it appears. The right-hand portion of Figure 10(a) shows a three phase mixture of γ, ε-grey, and (Nb)-white. In the central region, the microstructure consists of coupled (Nb)+γ eutectic surrounded by (Nb) dendrites and an (Nb)+ε matrix. The average composition of the central region was found to be 18.6 at.% Si, as shown by the compositional linescan in Figure 10(b), along line a-b in Figure 10(a).

Based on these analyses, a solidification sequence was deduced, as indicated on Figure 11. The figure shows the stable and metastable equilibria together with the experimentally determined (Nb)+ε coupled zone and the approximate boundaries of the metastable (Nb)+γ coupled zone. Referring to Figure 11, the solidification sequence may be summarized as follows: at the initial supercooling indicated by point "a" on the diagram, the alloy composition is outside the coupled zones. Thus, solidification begins with the formation of the primary (Nb) dendrites in the supercooled melt. The composition of the dendrites is given by the metastable extension of the (Nb) solidus, as indicated by point "b". The

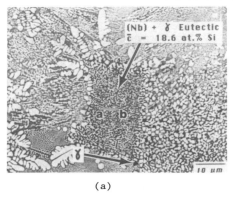

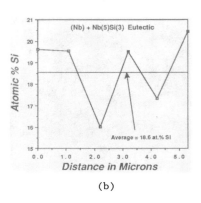

(a)   (b)

Figure 10 - Nb-17 at.% Si supercooled 190 K and splat-quenched: (a) BSE image of (Nb)+γ eutectic, and (b) corresponding linescan in the eutectic.

174

growth of the (Nb) dendrites with a composition < C_o causes a compositional shift in the remaining liquid toward higher Si contents, as well as a rise in temperature due to the recalescence effects. At some point during this process, the liquid composition will cross into the (Nb)+ε coupled zone as indicated by point "c". In this case, the (Nb)+ε eutectic is the preferred growth form and grows on the existing (Nb) dendrites, as seen in Figure 9(a). Note that the average composition of this eutectic will be slightly in excess of the initial alloy composition.

If, however, due to local solidification conditions, the (Nb)+ε eutectic is kinetically limited, further growth of (Nb) dendrites shifts the liquid composition into the coupled growth zone of (Nb)+γ eutectic. Thus, the metastable (Nb)+γ eutectic can nucleate and grow on existing (Nb) dendrites or the (Nb)+ε eutectic, as indicated by point "d". In this case, the liquid composition would be even richer in Si (due to a higher amount of (Nb) dendrites formed), which is indeed confirmed by the compositional analysis shown in Figure 10(b). Subsequent recalescence causes the liquid temperature to rise above the metastable eutectic isotherm, at which time the growth of (Nb)+γ eutectic must terminate. This may be one of the reasons why the (Nb)+γ eutectic regions are limited to a small size. The solidification of the remaining liquid can then occur first by (Nb)+ε coupled growth, followed by "normal" freezing, as indicated by point "e". Note that once the liquid enters the (Nb)+γ coupled zone, both the stable (Nb)+ε and the metastable (Nb)+γ eutectics can form. This was indeed observed, as indicated earlier in Figure 10(a).

The above analysis indicates the manner in which the relatively complex microstructural morphologies form depending on the nucleation and growth kinetics and relative domains of coupled zones for the stable and metastable eutectics.

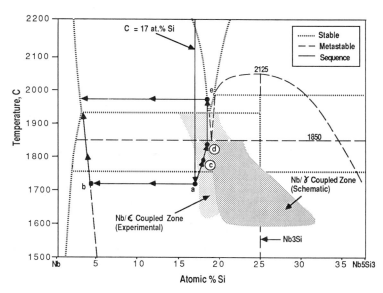

Figure 11 - Solidification sequence for Nb-17 at.% Si supercooled 190 K (see text).

(3) Suppression of γ in Nb-20 at.% Si. The microstructure of the alloy solidified in a "normal" manner (quenched from a superheated state) is shown in Figure 12. The back scattered electron micrograph shows properitectic γ (black) surrounded by ε (grey), in a (Nb)+ε eutectic matrix. The properitectic γ is not completely consumed because of the incomplete peritectic reaction. The microstructure of the same alloy solidified from a supercooling of 390 K is shown in Figure 13. The microstructure shows a fine coupled eutectic between (Nb) and ε, indicating suppression of the properitectic γ. Similar to the previous case, the observed microstructure is again compositionally and morphologically metastable.

Figure 12 - BSE micrograph of Nb-20 at.% Si quenched from a superheat of 90 K (black = γ; grey = ε; matrix = (Nb)+ε eutectic).

Figure 13 - BSE micrograph of Nb-20 at.% Si supercooled 390 K and quenched on a Cu-chill. Shows cells of fine (Nb)+ε eutectic.

(B) Metastable Phase Formation

The observed metastable phases were classified according to two categories as follows: (1) formation of a new crystalline phase, i.e., one whose existence cannot be derived (a priori) on the basis of metastable extensions of the equilibrium diagram and (2) glass formation.

(1) New Crystalline Phase Formation. In supercooled Nb 17 to 20 at.% Si alloys, single phase regions containing 19 to 20 at.% Si were detected in some samples. The morphological appearance of the phase was found to be dependent upon the original alloy composition, as illustrated in Figures 14 and 15(a) and 15(b). In each case, the samples were quenched on the Cu-chill plate. Figure 14 shows a plate or flake-like single phase region in a sample of Nb-17 at.% Si, solidified from a supercooling of 330 K. The composition of the plates was determined to be 19 at.% Si. The possibility that these single phase regions are a supersaturated stable phase must be ruled out because no metastable construction can be used to arrive at such a composition. Therefore, these single phase regions are presumed to be a new metastable phase.

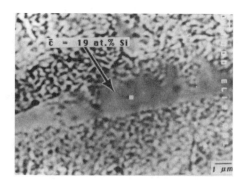

Figure 14 - Metastable phase in Nb-17 at.% Si supercooled 330 K and quenched on Cu-chill.

Figures 15(a) and 15(b) show another multicrystalline single phase region containing 20 at.% Si in the form of a rosette in a Nb-20 at.% Si sample solidified from a supercooling of 380 K. Figure 15(a) is a secondary electron image while Figure 15(b) is the corresponding back-scattered electron image. Further investigation of this morphology is currently in progress.

The measured composition of the single phase regions corresponds approximately to the stoichiometry of Nb_4Si which has been the subject of considerable controversy. Some have claimed Nb_4Si to be a stable phase [11,12], while others claimed that it does not exist as a stable phase [13,14]. However, the possibility that Nb_4Si may be a metastable phase has not been disproved.

(2) Glass Formation. During TEM analysis, amorphous regions were detected near the Cu-chill quenched surface of Nb-20 at.% Si samples supercooled by 250 K. The amorphous regions were typically small (≈ 1.0 μm) and surrounded by a crystalline two-phase mixture, as illustrated by

(a)   (b)

Figure 15 - Nb-20 at.% Si supercooled 380 K and quenched on Cu-chill (a) SE image of a metastable phase in the form of a rosette and (b) corresponding BSE image.

177

the bright field micrograph in Figure 16. The corresponding diffraction patterns are shown in Figure 17(a) to 17(c). The pattern for the amorphous region was obtained by CBED while those for the crystalline phases by SAD. Based on diffraction information, the two crystalline phases were identified as (Nb) (mottled contrast) and γ_1 (matrix; low temperature form of Nb_5Si_3).

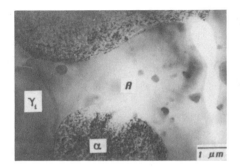

Figure 16 - TEM bright field image of amorphous region in Nb-20 at.% Si supercooled 250 K and quenched on Cu-chill: A-amorphous, γ_1-low temperature form of Nb_5Si_3, and α-(Nb).

(a)

Figure 17 - (a) CBED pattern from amorphous region and (b) SAD pattern for γ_1, and (c) SAD pattern for α.

(b)

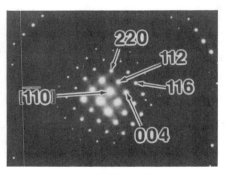

(c)

The fine scale of the amorphous region was revealed upon slight defocussing of the beam, which gave rise to faint diffraction spots from the surrounding crystalline matrix, as shown in Figure 18. It is interesting to note that the interface between the amorphous and crystalline regions is not well defined, as shown in Figure 19, which is a higher magnification bright field image of the interface area in Figure 16. The lack of a definite interface between either of the crystalline phases and the amorphous region indicates a gradual structural transition. Although the precise nature of this transition is unclear, it appears that devitrification of a portion of the amorphous regions has taken place at temperatures much lower than the solidification temperature.

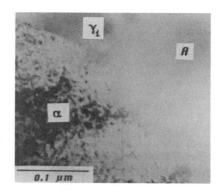

Figure 18 - CBED pattern from region A in Figure 17 under slightly defocussed beam conditions.

Figure 19 - TEM bright field image of the amorphous/crystalline interface in Figure 17.

There is another indication that this may in fact be the case: the relative volume fractions of (Nb) and γ_1 in the surrounding matrix were 36 and 64%, respectively. These values differ significantly from the lever rule proportions calculated from metastable extensions of the equilibria (46% (Nb) and 54% γ_1). The difference in these volume fractions can be reconciled only by assuming decomposition of the amorphous regions to yield non-equilibrium proportions of the two crystalline forms.

For the amorphous structure formation, it is important to prevent the nucleation and growth of the crystalline phases in any portion of the supercooled melt [15]. We have observed that the nucleation of crystalline phases is usually accompanied by a massive recalescence, which precludes the formation of amorphous structures. Thus, prior nucleation of crystalline phases must have been suppressed for the observed glass formation to occur. The glass formation for alloys near the (Nb)+ϵ eutectic composition can be predicted from an expression for the glass forming ability (GFA) proposed by Donald and Davies [16]. Their model for GFA is based upon the departure of the liquidus temperature from that calculated by rule-of-mixtures. Based on literature data, they concluded that GFA > 0.2 indicates a relative ease of glass formation. A plot of GFA vs. composition for Nb-rich Nb-Si, superimposed on the equilibrium diagram, is shown in Figure 20. The calculation indicates that alloys in the range 13 to 22 at.% Si may be promising candidates for the formation of amorphous structures.

179

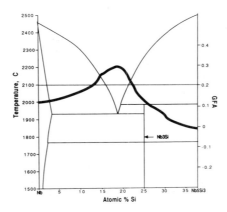

Figure 20 - Plot of Glass Forming Ability vs. composition.

Glass formation in these near eutectic alloys has been reported via melt spinning [17] and splat quenching [18-21], while Wang, et al. [22] extended the composition range of glass forming alloys via laser surface melting. However, we are unaware of any published reports of glass formation in these alloys via bulk supercooling.

Summary

The coupled zone for the (Nb)+ε eutectic was experimentally determined by analysis of a range of alloy compositions quenched from various degrees of bulk supercooling. The coupled zone was then used to deduce the solidification sequences of other processed alloys. A metastable (Nb)+γ eutectic was observed in some of the processed alloys. However, due to the infrequent appearance of this morphology, its coupled zone could not be determined experimentally.

A metastable crystalline phase with a composition of ~ 20 at.% Si was detected in alloys in the range of 17 to 20 at.% Si. This composition corresponds approximately to the stoichiometry of Nb_4Si. The phase is assumed to be a metastable phase since there is no possible metastable construction which can be used to (a priori) arrive at such a composition.

Calculation of the glass forming ability (GFA) for this system indicated that alloys near the Nb-rich eutectic may be relatively easy glass formers. Small amorphous regions were detected in bulk supercooled Nb-20 at.% Si samples. While the literature contains several reports of glass formation in these alloys via rapid quenching techniques such as melt spinning and laser quenching, the present results indicate that amorphous Nb-Si alloys can be formed via bulk supercooling.

Acknowledgement

This research was supported by the NASA-Vanderbilt Center for the Space Processing of Engineering Materials. Special thanks are due to Dr. Augusto Morrone for his assistance with the TEM analysis, and to John Haygarth and Dick Lewis for their valuable input.

180

References

1. A.B. Gokhale, G. Sarkar, G.J. Abbaschian, J.C. Haygarth, C. Wojcik, and R.E. Lewis, "Supercooling Effects in Faceted Eutectic Nb-Si Alloys," Solidification Processing of Eutectic Alloys, Ed.: by D.M. Stefanescu, G.J. Abbaschian, and R.J. Bayuzick, Publ: TMS Publications, 1988, 177-197.

2. K.R. Javed, A.B. Gokhale, and G.J. Abbaschian, "Rapid Solidification of Nb-based Alloys," Metals and Metals Processing (Proc. Conf.), Dayton, Ohio, August 1988.

3. J.A. Patchett and G.J. Abbaschian, "Grain Refinement of Copper by the Addition of Iron and By Electromagnetic Stirring," Met. Trans. B., 16B, (1985) 505-511.

4. A. Munitz and G.J. Abbaschian, "Effects of Supercooling on Solute Distribution and Microstructure," Undercooled Alloy Phases, Ed.: E.W. Collings and C.C. Koch, TMS Publications, 1987, 23-48.

5. A. Munitz and G.J. Abbaschian, "Solidification of Supercooled Fe-Ni Alloys," Advanced Materials and Manufacturing Processes, 3(3) (1988) 419-446.

6. S.P. Elder and G.J. Abbaschian, "Supercooling and Rapid Solidification Using Electromagnetic Levitation," to be published in the Proceedings of the Indo.-U.S. Conference on Solidification Processing, January, 1988.

7. D. Turnbull, "Metastable Structures in Metallurgy," Met. Trans. A, 12A (1981) 695-708.

8. W.J. Boettinger, D. Shechtman, R.J. Schaefer, and F.S. Biancaniello, "The Effect of Rapid Solidification Velocity on the Microstructure of Ag-Cu Alloys," Met. Trans. A, 15A (1984) 55-66.

9. R. Elliot, Eutectic Solidification Processing, Crystalline and Glassy Alloys, Butterworth and Co., Ltd., London, England, 1983, p. 291.

10. W. Kurz and D.J. Fisher, "Dendrite Growth in Eutectic Alloys: The Coupled Zone," Int. Met. Review, 5-6 (1979) 177-204.

11. A.G. Knapton, "The System Niobium-Silicon and the Effect of Carbon on the Structure of Certain Silicides," Nature, 175 (1955) 730.

12. R.P. Elliot, Constitution of Binary Alloys, First Supplement, McGraw-Hill, New York, NY, 1965, p. 270.

13. S.I. Alyamovskii, P.V. Geld, G.P. Shveikin, and I.I. Matveenko, "Existence of the Lower Nb Silicide, Nb_3Si," Inorg. Mater., 3(4) (1967) 649-652.

14. D.K. Deardorff, R.E. Siemens, P.A. Romans, and R.A. McCune, "New Tetragonal Compounds Nb_3Si and Ta_3Si," J. Less-Common Met., 18 (1969) 11-26.

15. D.H. St. John, "Freezing Diagrams: Part I. Development and Implications for Glass Formation," Met. Trans. A, 20A (1989) 287-299.

16. I.W. Donald and H.A. Davies, "Prediction of Glass Forming Ability for Metallic Systems," J. Non-Cryst. Solids, 30 (1978) 77-85.

17. M.T. Clapp and T. Manzur, "The Effect of Nitrogen Implantations on A15 Phase Formation in Nb_3Si," J. Appl. Phys., 61(2) (1987) 792-795.

18. R.M. Waterstrat, F. Haenssler, and J. Muller, "Nb-Si A15 Compounds By Liquid Quenching," J. Appl. Phys., 50(7) (1979) 4763-4766.

19. C. Suryarayana, W.K. Want, H. Iwasaki, and T. Masumoto, "High-Pressure Synthesis of A15 Nb_3Si Phase From Amorphous Nb-Si Alloys," Solid State Comm., 34 (1980) 861-863.

20. K. Togano, H. Kumakura, and K. Tachikawa, "Metastable A-15 Phase in Liquid Quenched Nb-Si Binary Alloys," Phys. Lett., 75A (1)(1980) 83-85.

21. W.K. Wang, Y. Syono, T. Goto, H. Iwasaki, A. Inone, and T. Masumoto, "Formation of FCC Solid Solution in Nb-Si System Under Shock Compression," Scripta Met., 15 (1981) 1313-1316.

22. W.K. Wang, C.J. Lobb, and F. Spaepen, "Formation of Metastable Nb-Si Phases By Picosecond and Nanosecond Pulsed Laser Quenching," Mat. Sci. and Eng., 98(1-2) (1988) 325-328.

ABSORPTION/DESORPTION BEHAVIOR OF NITROGEN

IN LIQUID NIOBIUM

H.G. Park, A.B. Gokhale, P. Kumar*, and G.J. Abbaschian

Department of Materials Science and Engineering
University of Florida
Gainesville, Florida 32611

*Cabot Corporation
Boyertown, Pennsylvania 19512

The absorption and desorption behavior of nitrogen in liquid Nb was investigated in temperature range 2740 to 2940 K under various nitrogen partial pressures using an electromagnetic levitation technique. The nitrogen solution reaction in liquid Nb was found to be exothermic, the standard enthalpy and entropy of solution being -236.4 kJ/mol and -65.3 J/K·mol, respectively. The absorption process is second-order with respect to nitrogen concentration, indicating that the rate controlling step is either the adsorption of nitrogen molecules on the liquid surface or dissociation of adsorbed nitrogen molecules into atoms. The desorption proces was found to be second-order as well.

Refractory Metals: State-of-the-Art 1988
Edited by P. Kumar and R.L. Ammon
The Minerals, Metals & Materials Society, 1989

Introduction

A unique combination of properties such as relatively low density, high melting point, and high superconducting transition temperature make Nb and its alloys attractive candidate materials in a variety of applications. The strength and electrical resistivity of these materials are known to depend strongly upon the amount of dissolved interstitial impurities such as C, N, O, and H. For example, the room temperature electrical resistivity of Nb increases linearly with nitrogen content, corresponding to a relative increase of 30% per atomic percent nitrogen dissolved [1].

Although the interstitials may be absorbed during service, especially at high temperatures, the major uptake of the interstitials takes place during melting and solidification. Efficient refining and control of the interstitial content during melting require a fundamental understanding of the interaction processes occurring between the gas phase and liquid Nb. Unfortunately, the high melting point of Nb (2468°C) precludes the use of a majority of the conventional experimental techniques used to study such processes. As a result, little fundamental data on the interaction of interstitials with liquid Nb are available at present.

In this paper we present our results on the absorption/ desorption behavior and solubility of nitrogen in liquid Nb. The data were obtained under various combinations of temperature, time, nitrogen partial pressure, and gas flow rate. Based upon the kinetic model, the results were analyzed in terms of the order of reaction and the rate controlling step.

All experiments were carried out in an electromagnetic levitation (EM) apparatus. EM levitation is particularly suited for the present study since this containerless processing method minimizes impurity pick-up from container walls. Also, the high melting temperature of Nb is not a serious constraint since heat generation takes place due to eddy current losses inside the sample. Furthermore, the system offers the advantages of rapid heating under various gas atmospheres, followed by rapid quenching if desired. The sample can be cycled through desired temperature ranges for long periods of time, thus offering a way to monitor the dissolution process as a function of time, temperature, and reactive gas partial pressure.

Experimental

Niobium rods of 0.25 inch diameter were used in this investigation. Compositional analysis of the as-received material is given in Table I. The experimental set-up is shown schematically in Figure 1. Details on EM levitation techniques are described elsewhere [2].

Table I: Chemical Analysis of Nb, ppm

Element	N	O	C	H	Ta	Al	Si	Fe	Nb
Concen-tration	15-30	18-125	15-25	5-10	395-790	10-15	0-10	10-25	Bal.

For studying the nitrogen absorption in liquid Nb, the following experimental procedure was used: first, a pre-cleaned cylindrical piece

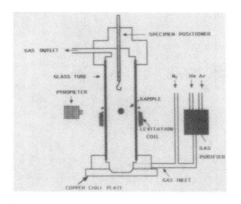

Figure 1

Experimental Set-up

of Nb weighing 2.1 ± 0.1 g was levitated in a mixture of purified Ar + He. The trace amount of oxygen in the inert gases was removed by passage through a column filled with titanium sponge held at 800°C. After the sample was stabilized in the levitated state, the He supply was cut off to raise the sample temperature, which was measured by a single color pyrometer calibrated against the melting temperature of pure Nb. The error in temperature measurement was ±5°C. Once the sample reached the reaction temperature, a premixed gas atmosphere with a desired ratio of N_2:Ar flowing at a predetermined rate was established almost instantaneously. This was taken as time zero for measuring the reaction time. After a desired reaction time, the power supply to the levitation coil was cut off and the sample was quenched on a copper chill plate. Simultaneously, the flow of Ar + N_2 was replaced with a Ar + He flow. In all experiments, the gas flow was measured using calibrated spherical float flowmeters. Using the procedure described above, the absorption behavior of nitrogen in liquid Nb was investigated under various conditions of nitrogen partial pressure, temperature, and time.

In studying the desorption behavior of nitrogen, the experimental procedure was modified as follows: first, the levitated sample was maintained at 2670°C and exposed to a stream of Ar + N_2 mixture (P_{N_2} = 0.015 atm) with total flow rate of 4000 cc/min for 4 minutes. As described later in the section on absorption kinetics, this treatment ensured that the Nb sample was saturated with nitrogen. Following this initial treatment, a new ratio of N_2:Ar was established in the reaction tube, again with a total flow rate of 4000 cc/min. The desorption kinetics were then monitored as a function of nitrogen partial pressure in the gas stream and the processing time. It is to be noted that in the present experimental setup, the Reynolds number for a flow rate of 4000 cc/min was found to be 480. Thus, the flow in the reaction chamber is expected to be laminar.

The nitrogen content of the processed samples was measured by two methods: (1) by recording the % weight gain/loss, and (2) by the LECO method. At large nitrogen concentrations, the latter method was found to be unreliable with non-systematic errors in the measured nitrogen content of replicate samples. For this reason, only the % weight gain/loss data have been used to analyze the interaction behavior. Considering that the vapor pressure of Nb at its melting point is ≈ 10^{-3} torr[3], the weight loss due to vaporization for 4 minutes under the experimental conditions

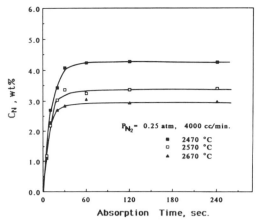

(a)

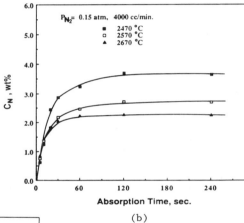

(b)

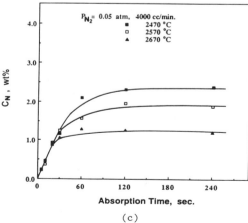

(c)

Figure 2 - Change in nitrogen concentration during nitrogen absorption into liquid Nb under various temperatures.
(a) at P_{N_2} = 0.25 atm,
(b) at P_{N_2} = 0.15 atm,
(c) at P_{N_2} = 0.05 atm

used is estimated to be ≈ 20 ppm. Therefore, it can be assumed that weight change in the processed samples is due to the interaction of nitrogen with liquid Nb. In other words, there is little difference between the saturated and equilibrium concentrations under the given experimental conditions.

Results

The effect of reaction temperature on absorption kinetics was investigated for three temperatures, 2470, 2570, and 2670°C. At each temperature, the absorption data were obtained for various lengths of time under a constant flow rate of 4000 cc/min of Ar + N_2 mixture with different nitrogen partial pressures. The results of nitrogen absorption experiments are summarized in Figure 2 which show the variation of nitrogen concentration with time for different processing temperatures at fixed nitrogen partial pressures of 0.25, 0.15, and 0.05 atm, respectively. Nitrogen absorption is rapid at the initial stage, while the absorption rate decreases with increasing processing time, the nitrogen concentration being saturated in 4 minutes in most cases. In each case, it is clear that the saturation concentration decreases with increasing processing temperature. This indicates that the process of nitrogen absorption in liquid Nb is exothermic.

In order to examine the effect of nitrogen partial pressure, the data obtained at 2570°C under various partial pressures of N_2 are plotted as a function of processing time in Figure 3. The increase in nitrogen partial pressure results in a saturation in shorter time and at a higher saturation level. This trend was also observed at two other processing temperatures (not shown).

Figure 4 shows the typical nitrogen desorption behavior for three nitrogen partial pressures at a reaction temperature of 2670°C. In this figure, the initial concentration is taken from the saturation concentration of initial absorption condition. It can be seen that lowering the nitrogen partial pressure in the gas stream increases the desorption rate.

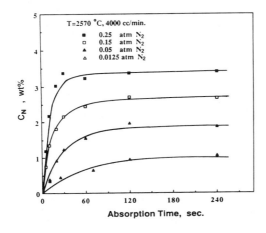

Figure 3 - Change in nitrogen concentration during nitrogen absorption into liquid Nb at 2570°C under 4 different nitrogen partial pressures.

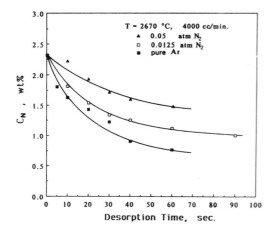

Figure 4 - Change in nitrogen concentration during desorption in liquid Nb at 2570°C.

Solubility of Nitrogen

The solution reaction of nitrogen in Nb may be described by the following equation:

$$\tfrac{1}{2} \, N_2(g) \;=\; N \text{ (in liquid Nb)} \tag{1}$$

the equilibrium constant, K, for the reaction is given by:

$$K = \frac{a_N}{\sqrt{P_{N_2}}} = \exp \, (-\Delta G^\circ / RT) \tag{2}$$

where a_N is the activity of N in liquid Nb at temperature T, P_{N_2} is the partial pressure of nitrogen in the gas, and ΔG° is the Gibbs energy change for the solution reaction. If the solution obeys Henry's law, i.e. if the activity-concentration relationship is linear, and hypothetical 1 wt.% solution of nitrogen is taken as the standard state, the activity term in Equation (2) can be directly replaced by wt.%. Consequently, Equation (2) becomes identical with Sieverts law, and the concentration of N in Nb can be related to the partial pressure of N_2 as follows:

$$C_N = \exp \, (\frac{\Delta S^\circ}{R} - \frac{\Delta H^\circ}{RT}) \cdot P_{N_2}^{\tfrac{1}{2}} = k(T) \cdot P_{N_2}^{\tfrac{1}{2}} \tag{3}$$

where C_N is the concentration of nitrogen in liquid Nb in wt.%, $k(T)$ is the temperature dependent Sieverts constant, and ΔH° and ΔS° are the standard enthalpy and entropy of solution, respectively. According to Equation (3), if the solution reaction is exthothermic, the solubility will decrease with increasing temperature and vice versa.

Based upon Equation (3), the saturated nitrogen concentration at each nitrogen partial pressure and reaction temperature is plotted as function as the square root of nitrogen partial pressure as shown in Figure 5. A marked deviation from linearity is observed at high nitrogen concentrations, indicating that Sieverts law does not hold over the entire range of temperatures and partial pressures investigated.

Nevertheless, Sieverts law can be applied at low nitrogen concentration (below approximately 2 wt.%). Taking Sieverts constant at each temperature, k, which is the slope of the linear portion of the plots in Figure 5, the values of log k are plotted as a function of 1/T in Figure 6. From the slope and intercept of this plot, the standard enthalpy and entropy of solution were calculated to be -236.4 kJ/mol and -65.3 J/K·mol, respectively. Using these quantities, the solubility of nitrogen in liquid Nb can be represented as a function of temperature and partial pressure of nitrogen as follows:

$$\log C_N = \frac{12356}{T} - 3.41 + \tfrac{1}{2} \log P_{N_2} \quad (< \approx 2 \text{ wt.%}) \tag{4}$$

where C_N is in wt.%, T is in K, and P_{N_2} is in atm. It must be emphasized that Equation (4) is valid only in the range where Sieverts law is applicable.

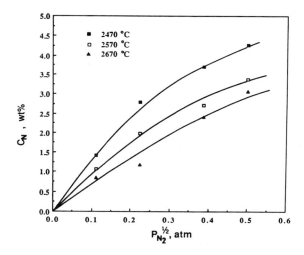

Figure 5 - Plot of equilibrium nitrogen concentration as a function of $P_{N_2}^{1/2}$ at various temperatures.

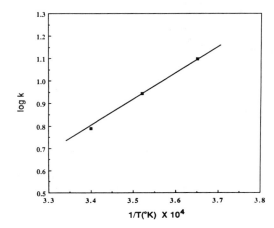

Figure 6 - Arrhenius type plot of log k, Sievert constant, versus 1/T.

Kinetics

Since nitrogen absorption in liquid Nb is a heterogenous process, it can be considered to occur in the following five distinct elementary steps, shown schematically in Figure 7:

  (i)   transport of nitrogen molecules from the gas phase to the gas/metal interface,

  (ii)   adsorption of nitrogen molecules on the metal surface,

  (iii)   dissociation of the adsorbed molecules into adsorbed atoms,

  (iv)   incorporation of the adsorbed atoms to subsurface layers, and

  (v)   transport of nitrogen atoms to bulk metal.

In the desorption reaction the same steps occur in the reverse sequence.

If step (i) is the rate controlling step, the rate of nitrogen absorption would be influenced by the flowrate of the input gas stream because the gas phase mass transfer coefficient is a function of Reynold number which includes a gas-velocity term. As reported elsewhere [4], the absorption rate was affected by the gas flow rate in the lower total flow rate. However, beyond a certain value of flow rate, it becomes independent of gas flow rate. Since the gas flow rate of 4000 cc/min, which is maintained in all experiments, was in the regime such that gas phase transport is not a rate controlling step, the step (i) will not be taken into consideration in determining the rate controlling step.

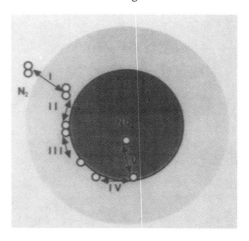

Figure 7 - Mechanism of nitrogen absorption/desorption in liquid Nb:
  I   Gas phase transport
  II Adsorption of nitrogen molecules on the surface
  III Dissociation of nitrogen molecules into atoms
  IV Incorporation of nitrogen atoms into subsurface layer
  V   Liquid phase transport

In case that step (v) is a rate controlling step, any change in hydrodynamics of the liquid phase will have a considerable effect on the rate of nitrogen absorption. It is likely that vigorous convective motion present in the levitated liquid Nb droplet under electromagnetic field will greatly increase the mass transfer rate. This is supported by Lee and Rao's experiment[5] on the diffisivity of carbon in an iron-carbon droplet where the carbon mass transfer in the liquid phase was enhanced by about ten times under electromagnetic conditions compared to a stationary condition. Consequently, step (v) is not expected to be a rate controlling step in the experimental system used in this investigation.

In determining the rate controlling step among other three elementary steps, it is common to compare the data obtained experimentally with theoretical rate law. Through the mathematical treatment under the assumption that one of those controls the overall reaction dominantly, it has been reported by earlier researchers [6-8] that if step (ii) or (iii) is the rate limiting step, then the reaction is a second-order process with respect to nitrogen concentration, while if step (iv) controls the overall reaction rate, then the reaction is a first-order process. The two types of process can be represented as follows:

First-order reaction:

$$\frac{dC_N}{dt} = k_1 \left(\frac{A}{V}\right) (C_{Ne} - C_N) \tag{5a}$$

or

$$-\ln \left(\frac{C_{Ne} - C_N}{C_{Ne} - C_{No}}\right) = k_1 \left(\frac{A}{V}\right) t \tag{5b}$$

Second-order reaction:

$$\frac{dC_N}{dt} = k_2 \left(\frac{A}{V}\right) (C_{Ne}^2 - C_N^2) \tag{6a}$$

or

$$\frac{1}{2C_{Ne}} \left[\ln \left(\frac{C_{Ne} + C_N}{C_{Ne} - C_N}\right) + \ln \left(\frac{C_{Ne} - C_{No}}{C_{Ne} + C_{No}}\right)\right] = k_2 \left(\frac{A}{V}\right) t \tag{6b}$$

where C_N, C_{No}, and C_{Ne} are nitrogen concentration at time t, at t=0, and at equilibrium, respectively. A is specimen surface area, V is the melt volume, and k_1 and k_2 are the first- and second-order reaction constant, respectively. It should be noted that the rate constants k_1 and k_2 are a function of temperature only. In order to determine if the overall reaction is first-order or not, the absorption data collected at 2470, 2570, and 2670°C under various partial pressures of nitrogen are plotted in Figure 8 using Equation (5b). In these figures, the slopes indicate the rate constant, k_1. It can be seen that in each case, the data are split into three or four distinct lines, each representing a different partial pressure of N_2 in the gas mixture. If the overall reaction is first-order, the data obtained at a given temperature should fall on one line irrespective of the nitrogen partial pressure because the rate constant is determined by reaction temperature only. Accordingly, this finding suggests that the rate controlling step in nitrogen absorption in liquid Nb is not first-order. This implies that the elementary step (iv) does not control the overall reaction.

191

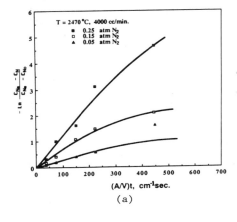

(a)

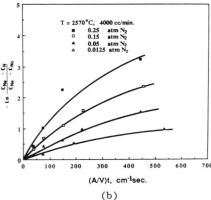

(b)

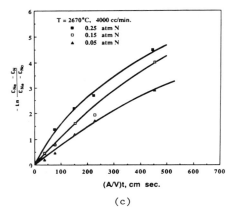

(c)

Figure 8 - Kinetics of nitrogen absorption into liquid Nb under various nitrogen partial pressures correlated with first order rate equation (a) at 2470°C, (b) at 2570°C, and (c) at 2670°C.

In order to examine the possibility of a second-order reaction, the left hand side term in Equation (6b) was plotted as a function of $(A/v)t$ for each reaction temperature. This is shown in Figure 9 for temperatures of 2470, 2570, and 2670°C, respectively. It can be seen that within the limits of experimental error, the data in each case fall on a single straight line regardless of the N_2 partial pressure used. Thus, it is concluded that either elementary step (ii) or (iii), which are both second-order, must be controlling the overall reaction rate.

Since nitrogen desorption is expected to occur in reverse sequence, the data analysis is exactly analogous. The desorption data in Figure 4 are plotted in accordance with Equations (5b) and (6b) in Figures 10 and 11, respectively. In these figures, the initial concentration, C_{No}, and the equilibrium concentration, C_{Ne}, were taken from the absorption data obtained under the same processing conditions. The first-order plot shows a splitting similar to that in absorption, while the data correlate well with the second-order plot.

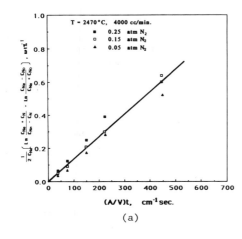

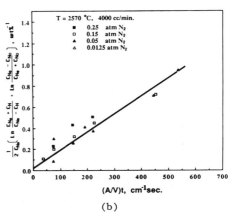

(a)

(b)

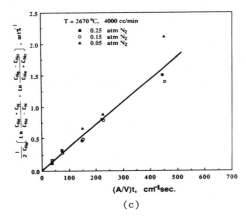

(c)

Figure 9 - Kinetics of nitrogen
absorption into liquid Nb under
various nitrogen partial
pressures correlated with
second-order rate equation.
(a) 2470°C, (b) 2570°C, and (c)
2670°C.

Based on this analysis, we conclude that the desorption reaction is again controlled by elementary steps (ii) or (iii).

Summary

Using an electromagnetic levitation technique, absorption and desorption behavior of nitrogen in liquid niobium was investigated under various experimental conditions of temperature, nitrogen partial pressure and gas flow rate. Some important results are as follows:

1. Nitrogen solution reaction in molten niobium is exothermic. Standard solution enthalpy and entropy are -236.4 kJ/mol and -65.3 J/K·mol, respectively. The solubility of nitrogen can be represented as a function of nitrogen partial pressure and temperature as follows:

$$\log C_N = \frac{12356}{T} - 3.41 + \tfrac{1}{2} \log P_{N_2} \qquad (< \approx 2 \text{ wt.\%})$$

where C_N is in wt.%, T is in K, and P_{N_2} is in atm.

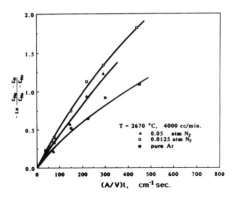

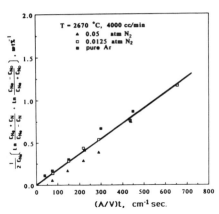

Figure 10 - Kinetics of nitrogen desorption in liquid Nb at 2670°C correlated with first-order reaction equation.

Figure 11 - Kinetics of nitrogen desorption in liquid Nb at 2670°C correlated with second-order rate equation.

2. Both absorption and desorption of nitrogen in molten niobium are second-order processes with respect to nitrogen concentration. The rate controlling step in absorption process is expected to be either the adsorption of nitrogen molecules on the liquid surface or the dissociation of nitrogen molecules into nitrogen atoms.

Acknowledgment

This research was supported by NASA through the NASA-Vanderbilt Center for the Space Processing of Engineering Materials.

References

1. E. Fromm and G. Horz: Intl. Met. Rev., no. 5-6, 1980, p. 269.

2. R.I. Asfanani, D.S. Shahapurkar, Y.V. Murty, J.A. Patchett, and G.J. Abbaschian: J. of Metals, Vol. 37, No. 4, 1985, p. 22.

3. J.F. O'hanlon: A User's Guide to Vacuum Technology, p. 372, John Wiley & Sons, Inc., New York, 1980.

4. H.G. Park, A.B. Gokhale, P. Kumar, and G.J. Abbaschian, submitted for publication.

5. H.G. Lee and Y.K. Rao: Met. Trans. B, vol. 13B, September, 1982, p. 411.

6. H.G. Lee and Y.K. Rao: Ironmaking Steelmaking, vol. 12, 1985, p. 221.

7. K. Mori, M. Sano, and K. Suzuki: Trans. ISIJ, vol. 13, no. 1, 1973, p. 60.

8. Y.K. Rao and H.G. Lee: I.S.S. Trans., vol. 4, 1984, p. 1.

Subject Index

▶

Author Index